Student Workbook to Accompany

Robert S. Witte

Statistics

THIRD EDITION

Prepared by
GLENDA STREETMAN SMITH, Ph.D

Holt, Rinehart and Winston, Inc.
Fort Worth   Chicago   San Francisco   Philadelphia
Montreal   Toronto   London   Sydney   Tokyo

ISBN 0-03-014152-4

Copyright © 1989 by Holt, Rinehart and Winston, Inc.

Requests for permission to make copies of any part of the work should be mailed to:
Copyrights and Permissions Department
Holt, Rinehart and Winston, Inc.
Orlando, Florida 32887

Printed in the United States of America
9 0 1 2 018 9 8 7 6 5 4 3

Holt, Rinehart and Winston, Inc.
The Dryden Press
Saunders College Publishing

PREFACE

This workbook is designed to help you understand and learn the statistical concepts presented in the Witte Statistics, 3e.

The first section in each chapter is Learning Objectives. This should give you a clear idea of what you should be able to understand or accomplish when you have studied the chapter well.

The second section in the chapter is Key Terms. Learning the language of statistics is such an important part of a basic course that the significant vocabulary terms defined in the text are repeated here for further emphasis and exposure. Study these terms well.

The third section in each chapter is a Text Review. This section does exactly as its name implies, reviews the most important ideas in the chapter. The blanks that you fill in are there for the purpose of changing the reading/studying experience from a passive one to a more active one. Learning theory tells us that learning is more efficient and effective when the learner is active. Try to fill in each blank as you read. Then check your answers at the end of each chapter. Even if you get the answer wrong, the activity will be aiding your learning in the long run. The numbers in parentheses before the blanks are the ordinal numbers of the answer. The numbers in parentheses after the blanks refer to the sections of the text where the answers can be found.

The fourth section in each chapter consists of Problems and Exercises. Some of these are similar to those at the end of the chapter in your text. Others will provide somewhat different practice opportunities. Either way, additional practice with these problems and exercises will be a great way to check how much you have learned and to reinforce the learning.

The fifth section, Beyond the Basics, offers suggestions for advanced students who have mastered the concepts of the chapter and want to explore the ideas further without just working more of the same problems. Those who are really interested in pursuing some of the ideas presented in the text and learning more about the "real world" application of statistics should find this section particularly rewarding. Occasionally this section will contain something just a little different, a hint for how to study particular concepts or a unique presentation of the material.

The sixth section is a <u>Post Test</u> to check your understanding of the information presented in the Chapter. If you do well, you will know that you have mastered some of the most important concepts presented in that chapter. If not, you will know that you need more study time.

The final section of each chapter contains the answers to the text review, problems and exercises, and post test sections. The answers to many of the problems were computed using a computer. If you solve them with a hand calculator, your answer may vary slightly. One way to match more closely is to avoid rounding off at any phase of the problem until the answer is computed. Small differences in rounding during the many steps of some of the more complex calculations can ultimately result in a large discrepancy before the answer is found.

One last word about the problems is necessary. In order to facilitate ease of computation, sample sizes are unrealistically small in many of the problems presented for you to solve. Most of the research examples are made up and not necessarily based on real research. If you are interested in some of the ideas presented, you should do a library search to determine whether any real research has been conducted to investigate the problem of interest.

CONTENTS

CHAPTER 1

INTRODUCTION

Learning Objectives

You should develop knowledge and understanding of key terms.

Given a data set, you will be able to differentiate between quantitative and qualitative data.

Given research situations, you will be able to differentiate between descriptive and inferential statistics.

Key Terms

Data -- A collection of observations from a survey or experiment.

Descriptive statistics -- The area of statistics concerned with organizing and summarizing information about a collection of actual observations.

Inferential statistics -- The area of statistics concerned with generalizing beyond actual observations.

Quantitative data -- A set of observations where any single observation is a number that represents an amount or count.

Qualitative data -- A set of observations where any single observation is a word or code that represents a class or category.

Text Review

A researcher gathers information in the form of individual records called observations. These records could be IQ scores, height, weight, gas mileage of a particular car, or length of lifespan of white rats. A collection of observations is referred to as (1)_____ (1.4). In a set of observations, when any single record is a number that represents an amount or count, the data are (2)_____ (1.5). On the other hand, if a single observation is a word or code that represents

1

a class or category, the data are (3)_____
(1.6). Numbers may be arbitrarily assigned to survey
responses such as yes and no replies in order to
facilitate computer processing. Numbers may also appear
in certain instances where they represent only category
or classification. An example would be the number on a
player's football jersey. Number 88 simply designates
that this player is probably a tight end. It does not
indicate that he is twice as good a player as number 44,
who is probably a running back. When numbers are used in
this way, they represent (4)_____ data (1.7).

When a researcher wants to organize or summarize
information about a collection of observations, the older
area of statistics, (5)_____ (1.2) would be
used. When the researcher must generalize beyond the
actual observations, then (6)_____ (1.2)
would be used.

Problems and Exercises

Indicate whether the following data are quantitative or
qualitative.

1. social security numbers   _____
2. age   _____
3. weight   _____
4. socioeconomic status   _____
5. eye color   _____
6. years work experience   _____
7. movie ratings (G, PG, R, X) _____
8. top ten best sellers   _____
9. scores on statistics exam   _____
10. average amount of allowance
    for ten-year-olds   _____

Indicate whether the following situations would require
descriptive (D), or inferential (I), statistics.

1. D I   analysis of the 1990 census
2. D I   calculation of grade point average
3. D I   Neilsen ratings for a particular TV show
4. D I   "best team" performance award at a swim meet
5. D I   survey results from a magazine poll
6. D I   average age of students in your statistics
           class
7. D I   dentists' preference for brand of toothpaste

8.	D	I	number of football games won by your team
9.	D	I	average monthly percent inflation rate
10.	D	I	finance charge on your charge account

Beyond the Basics

In the library, locate some issues of a professional journal which reflect your interest. Look at the methods and results sections of the articles. Determine whether the statistical analysis was descriptive or inferential. Think about the reasons for doing an inferential analysis rather than a descriptive one. Are the data used quantitative or qualitative?

Post Test

1. The two main subdivisions of statistics are
   _____ and _____ .

2. _____ is used to organize and summarize data.

3. _____ is used to generalize beyond the actual observations.

4. A collection of observations from a survey or experiment is called _____ .

5. A set of observations where any single observation is a number that represents count or amount is _____ data.

6. A set of observations where any single observation is a word or code that represents a class or category is _____ data.

Answers

Text Review

1. data
2. quantitative
3. qualitative
4. qualitative
5. descriptive
6. inferential

Problems and Exercises

1. qualitative
2. quantitative
3. quantitative
4. qualitative
5. qualitative
6. quantitative
7. qualitative
8. qualitative (note: position on best seller list is a class or category, not number of books sold)
9. quantitative
10. quantitative

Descriptive or inferential statistics

1. D
2. D
3. I
4. D
5. I
6. D
7. I
8. D
9. I
10. D

Post Test

1. descriptive, inferential
2. descriptive
3. inferential
4. data
5. quantitative
6. qualitative

CHAPTER 2

DESCRIPTIVE STATISTICS:
ORGANIZING AND SUMMARIZING DATA

Learning Objectives

You should develop knowledge and understanding of key terms.

Given a data set, you will be able to construct and interpret a frequency distribution for grouped or ungrouped data.

Key Terms

Frequency distribution -- A collection of observations produced by sorting observations into classes that show their frequencies of occurrence.

Frequency distribution for ungrouped data -- A frequency distribution produced whenever observations are sorted into classes of single values.

Frequency distribution for grouped data -- A frequency distribution produced whenever observations are sorted into classes of more than one value.

Cumulative frequency distribution -- A frequency distribution showing the total number of observations at or below each class.

Relative frequency distribution -- A frequency distribution showing the frequency of each class as a part or fraction of the total frequency for the entire distribution.

Percentile rank of an observation -- Percentage of observations in the entire distribution with similar or smaller values than that observation.

Outlier -- A very extreme observation.

Unit of measurement -- The smallest possible difference between scores in a particular data set.

Text Review

There is no single right way to organize data, but a few guidelines make using tables and graphs relatively simple. A frequency distribution for ungrouped data results when observations are sorted into classes of (1)_____ (2.1) values. Observations are arranged in a column with the (2)_____ (2.1) observation at the bottom and the (3)_____(2.1) at the top. When the number of possible values is larger than twenty, a frequency distribution for (4)_____ (2.2) data is more appropriate. In this type of distribution, observations are sorted into classes of more than one value.

There are several guidelines for preparing frequency distributions. One is that each observation should be included in only (5)_____ (2.3) class. The second guideline is that all classes must be listed, even those with zero frequencies. The third guideline states that all classes must be (6)_____ (2.3) in width. The fourth guideline is optional and suggests that all classes must have both upper and lower (7)_____ (2.3). The fifth guideline instructs the researcher to select width of classes from (8)_____ (2.3) numbers. The sixth states that the lower boundary of each class should be a multiple of class width. The seventh guideline indicates that approximately (9)_____ (2.3) classes is the ideal number. However, this number should be flexible. Larger data sets might require more classes so that important data patterns are not overlooked. Smaller data sets may be more clearly portrayed with fewer classes.

In a well-constructed frequency table, each observation should be clearly assigned to one and only one class and the gaps between classes should always equal one (10)_____ (2.5).

A (11)_____ (2.7) frequency distribution shows the frequency of each class as a part or fraction of the total frequency for the entire distribution. This type of distribution helps to compare two distributions based on different numbers of observations. It can also be used to examine the relative concentration of observations among different classes within the same distribution.

A frequency distribution showing the total number of observations at or below each class is a (12)_____ distribution (2.9). The most effective use of this type of distribution is with data where relative standing within the distribution is important. A good example would be achievement or aptitude test scores. These cumulative percents are referred to as (13)_____ (2.10) when they describe the relative position of an observation within the distribution. When obtained from a distribution of ungrouped data, percentile ranks are exact, but when obtained from a distribution of grouped data, they are only (14)_____ (2.10).

Frequency distributions may be constructed for qualitative data. Note that the numerical summary of qualitative data in this way does not change it to quantitative data. Remember in Chapter 1 (p. 7, text), the emphasis was on a <u>single observation</u> when determining whether a set of observations represented qualitative or quantitative data. When qualitative data can be ranked, the data should be listed in (15)_____ order in the frequency table (2.12). The ranking of qualitative data makes it possible to construct cumulative frequency distributions, but (16)_____ (2.12) distributions can be constructed for qualitative data even when it cannot be ranked.

Occasionally an extreme observation will appear in a data set. Such extreme observations, called (17)_____ (2.13) should always be checked for accuracy.

Problems and Exercises

Guidelines for Frequency Distributions

1. Each observation should be included in one, and only one, class.
2. List all classes, even those with zero frequencies.
3. All classes (with upper and lower boundaries) should be equal in width.
4. All classes should have both an upper and a lower boundary.
5. Select the width of classes from convenient numbers.
6. The lower boundary of each class should be a multiple of the class width.
7. In general, aim for a total of approximately ten classes.

How to Construct a Frequency Distribution

1. Find the range, that is, the difference between the largest and smallest observation.
2. Find the class width required to span the data range by dividing the range by the desired number of classes.
3. Round off to the nearest convenient width.
4. Determine where the lowest class should begin.
5. Determine where the lowest class should end by adding the class width to the lower boundary and then subtracting one unit of measurement.
6. Working upward, list as many equivalent classes as are required to include the largest observation.
7. Indicate with a tally the class in which each observation falls.
8. Replace the tally count for each class with a frequency.
9. Supply headings for both columns and a title for the table.

---

1. On a 50-point exam for a General Psychology class, a professor observes the following results. Construct a frequency distribution showing relative and cumulative frequency.

48	37	22	38	39
45	42	29	48	33
29	41	40	34	30
39	38	42	44	32
31	39	35	40	36
42	30	33	41	46

Table 2.1

A grades	B f	C relative f	D cumulative f	E cumulative %

2. Which class interval would contain the score closest to the 50th percentile?

3. Which is appropriate for this data set, a grouped or ungrouped distribution? Why?

4. Describe any interesting patterns? In general, will the professor be pleased with this overall class performance?

5. What is the unit of measure for this data set?

6. A third grader's reading achievement test score is reported to his mother as a score at the 73rd percentile. What is the meaning of this score? If his math score is at the 66th percentile, relatively speaking, is he better in math or reading?

Beyond the Basics

Obtain your own data set by asking 30 people to count the change in their purse or pocket for you. Record their responses and construct a frequency distribution. Note any patterns. Were there any outliers present? If so, can you explain them.

Post Test

1. When observations are sorted into classes of single values, the result is referred to _____.

2. A frequency distribution for grouped data should be used when there are more than _____ possible values.

3. A problem sometimes encountered when data are grouped is _____.

4. When too many classes are used, a problem sometimes encountered is _____.

5. The size of the gap between classes should always equal _____.

6. _____ distributions can be especially useful when comparing the shapes of two or more distributions with unequal numbers of observations.

7. _____ indicates the percentage of observations in the distribution with similar or smaller values.

8. Cumulative frequency distributions can only be constructed for qualitative data if_____.

9. A very extreme observation is called _____.

Answers

Text Review

1. single
2. smallest
3. largest
4. grouped
5. one
6. equal
7. boundaries
8. convenient
9. 10
10. unit of measurement
11. relative
12. cumulative frequency
13. percentile rank
14. approximate
15. descending
16. relative frequency
17. outlier

Problems and Exercises

1.

Table 2.1

A grades	B f	C relative f	D cumulative f	E cumulative %
48-50	2	.07	30	100
45-47	2	.07	28	93
42-44	4	.13	26	87
39-41	7	.23	22	73
36-38	4	.13	15	50
33-35	4	.13	11	37
30-32	4	.13	7	23
27-29	2	.07	3	10
24-26	0	.00	1	3
21-23	1	.03	1	3

2. the interval of 36-38

3. Grouped, there are more than twenty possible values.

4. There is one outlier, the score of 22. In general,
    the most frequencies occurred in the interval of
    39-41. This would be a percentage correct of about
    78 to 82 on the test. However, over one third,
    eleven people, scored below 35, which would be 70
    percent correct. If 70 percent is passing, perhaps
    the professor would not be pleased that only 19 out
    of 30 students passed the test. She might also like
    to see more high grades in the intervals of 45-47
    and 48-50.

5. The unit of measurement is one score point.

6. The score at the 73rd percentile means that of all
    the people who took the test, this boy scored higher
    than or equal to 73 percent of them. He is better
    in reading than in math when compared to his peers.

Post Test

1. a frequency distribution for ungrouped data
2. twenty
3. information about each observation is lost
4. the distribution fails to provide a concise
    description of the data
5. the unit of measurement
6. relative frequency
7. percentile rank
8. the data can be ranked or ordered
9. an outlier

DESCRIBING DATA WITH GRAPHS

Learning Objectives

You should develop knowledge and understanding of key terms.

You will be able to construct and interpret histograms and bar graphs.

You will be able to recognize skewness in a distribution and determine the direction.

Key Terms

Histogram -- Bar-type graph for quantitative data. No gaps between adjacent bars.

Frequency polygon -- Line graph for quantitative data.

Stem and leaf display -- A device for sorting quantitative data on the basis of leading and trailing digits.

Positively skewed distribution -- A distribution that includes a few extreme observations with relatively large values in the positive direction.

Negatively skewed distribution -- A distribution that includes a few extreme observations with relatively small values in the negative direction.

Bar graph -- Bar-type graph for qualitative data. Gaps between adjacent bars.

Text Review

Making graphs from (1)_____ (3.0) helps to describe data more clearly. A bar-type graph appropriate for quantitative data is called a (2)_____(3.1). This graph is characterized by (3)_____ (3.1) units along the horizontal axis which indicate (4)_____ (3.1). The equal units along the vertical axis indicate (5)_____ (3.1).

Where the two axes meet, both frequency and class interval equal (6)_____(3.1). Along the horizontal axis, values always (7)_____ (3.1) from left to right.

The line graph, or (8)_____ (3.2), is a variation of the histogram which is particularly useful when (9)_____ (3.2).

Finally, quantitative data can be summarized using a (10)_____(3.3), which is somewhat of a cross between a (11)_____ and a (12)_____ (3.3). The advantage of the stem and leaf display is in summarizing data without losing information from the (13)_____ (3.3).

(14)_____ is an important characteristic of a frequency distribution, no matter how it is presented. The normal shape, which looks like a (15)_____ (3.4), reflects the actual distribution of many familiar data sets. Standardized test scores are usually represented by the normal curve. The (16)_____ (3.4) shape, with two humps, reflects the presence of two different types of observations in the same distribution. A lopsided shape with many observations to the left indicates the presence of many smaller values and only a few large values. This is referred to as (17)_____ (3.4) skewed. A lopsided shape with many observations to the right indicates the presence of many larger values and only a few small values. This is referred to as (18)_____ (3.4) skewed. Thus, skewness, unlike most other concepts, is defined by the (19)_____ (3.4) of observations, not the majority.

Qualitative data can be depicted more clearly using a (20)_____(3.5). The horizontal axis reflects (21)_____ and the vertical axis (22)_____ (3.5), just as in a histogram. Unlike the histogram, however, the bar graph has (23)_____ (3.5) between the bars, indicating the discontinuous nature of qualitative data.

Misleading graphs can be constructed to support a particular point of view. This is usually done by violating customary procedure. The correct method of constructing a graph is to make the horizontal and vertical axes (24)_____ (3.6) in height and width. The bars on the graph must also be (25)_____ (3.6) in width. And the entire scale must be reproduced

Problems and Exercises

Constructing Graphs

1. Decide on the appropriate type of graph.
2. Use a ruler to draw the horizontal and vertical axes (equal width and heigth).
3. Identify the string of class intervals that eventually will be superimposed on the horizontal axis.
4. Superimpose the string of class intervals (with gaps for bar graphs) along the entire length of the horizontal axis.
5. Along the entire length of the vertical axis, superimpose a progression of convenient numbers beginning with zero.
6. Using the scaled axes, construct bars to reflect the frequency of observations within each class interval.
7. Supply labels for both axes and a title for the graph.

---

1. A market research firm is hired to compare preferences of housewives as to comfort and ease of driving four different vehicles. The following results are obtained:

Vehicle	Number of Housewives Preferring
Four-wheel drive	36
Station Wagon	44
Sports car	24
Pick-up truck	13

Construct a bar graph describing these results.

2. The Keebler elves are being evaluated on a ten-point scale for cookie-packing aptitude. The following ratings are observed:

10	4	3	6	5	6	1	6	7	7
6	9	2	8	5	3	6	9	8	4
9	5	1	5	2	1	3	8	3	7

a. Construct a histogram for the above data.
b. Construct a frequency polygon.

3. In a statistics class at a community college, approximately half the students are nontraditional (about age 30 or older) and the other half traditional (age 19 or 20). The students earn the following scores on the first exam:

Nontraditional Students					Traditional Students				
69	68	77	94	76	67	87	98	78	77
88	76	75	79	69	83	84	87	77	90
79	66	98	72	83	78	76	78	70	72
81	81	82	83	92	74	75	66	70	70
86	92	77	82	71	68	66	84	66	67

16

a. Using the preceding data, construct two frequency polygons on the same graph showing the test results achieved by the two groups of students.
b. What is the shape of the distribution for nontraditional students?
c. What is the shape of the distribution for traditional students?
d. Overall, is the performance of the two groups similar or dissimilar?

Beyond the Basics

Look for histograms, bar graphs, or frequency polygons in current newspapers or magazines. Are there any features intended to be misleading? How might the use of computers have affected the presentation of misleading graphs?

Post Test

1. What determines whether to construct a bar graph or histogram for a given data set?

2. List three ways to construct graphs so that they are misleading.

3. What determines whether skewness of a distribution is positive or negative?

4. When is it beneficial to construct a frequency polygon instead of a histogram?

5. Why does a bar graph have gaps between the bars when a histogram does not?

6. What is the biggest advantage of a stem and leaf display?

Answers

Text Review

1. frequency distributions
2. histogram
3. equal
4. class intervals
5. increases in frequency
6. zero
7. increase
8. frequency polygon
9. comparing distributions
10. stem and leaf display
11. frequency table
12. histogram
13. raw scores
14. shape
15. bell
16. bimodal
17. positively
18. negatively
19. minority
20. bar graph
21. different words/classes
22. frequency
23. gaps
24. about equal
25. about equal
26. zero

Problems

1.

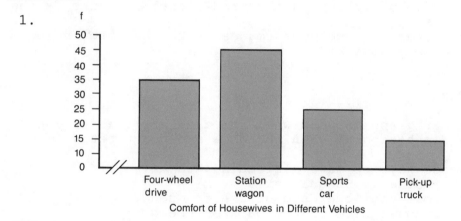

Comfort of Housewives in Different Vehicles

2.

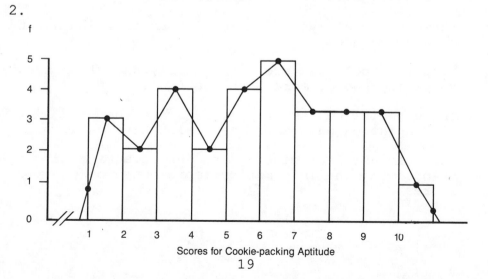

Scores for Cookie-packing Aptitude

19

3.

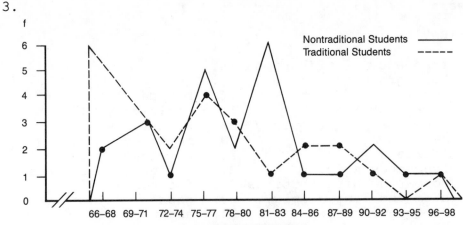

Test Results of Traditional
vs. Nontraditional Students

b.   The distribution for nontraditional students is
somewhat negatively skewed.
c.   The distribution for traditional students is the
same.
d.   Overall, the performance of the two groups is
very similar.

Post Test

1.   A bar graph is used when the data are qualitative
and a histogram when the data are quantitative.

2.   Three ways to make misleading graphs:
a. radically unequal axes
b. omit the lower end of the frequency scale
c. unequal size of the bars on the graph

3.   Skewness is always in the direction of the minority
of observations.

4.   Frequency polygons are particularly useful when
comparing two or more distributions.

5.   The gaps in the bar graph illustrate the
discontinuous nature of the qualitative data.

6.   The stem and leaf display has the advantage of
maintaining the information represented by the raw
scores.

DESCRIBING DATA WITH AVERAGES

Learning Objectives

You should develop knowledge and understanding of key terms.

You will be able to calculate and interpret the various measures of central tendency.

You will be able to determine which measure of central tendency is appropriate for use with a specific data set.

Key Terms

Measures of central tendency -- General term for the various averages.

Mode -- The value of the most frequent observation.

Mean ($\widehat{X}$) -- The balance point for a frequency distribution, found by dividing the total of all observations by the number of observations.

Median -- The middle value when observations are ordered from least to most.

Bimodal -- Relating to any distribution with two obvious peaks.

Text Review

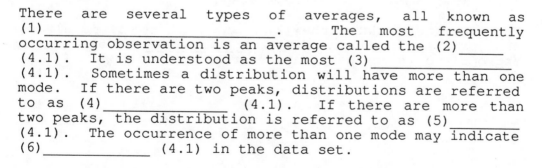

There are several types of averages, all known as (1)_____.   The most frequently occurring observation is an average called the (2)_____ (4.1). It is understood as the most (3)_____ (4.1). Sometimes a distribution will have more than one mode. If there are two peaks, distributions are referred to as (4)_____ (4.1). If there are more than two peaks, the distribution is referred to as (5)_____ (4.1). The occurrence of more than one mode may indicate (6)_____ (4.1) in the data set.

When observations are ordered from least to most, the middle value is the (7)_____ (4.2). The median has a percentile rank of (8)_____.(4.2) The median reflects the (9)_____ (4.2) of an observation, not the position.

The most common of the averages is the (10)_____ (4.3). Adding all the observations and then dividing by the number of observations will yield the (11)_____ (4.3). This calculation can be done without ordering the data or organizing it. When symbols are used, the number of observations is represented by (12)_____(4.3). The symbol for the mean is (13)_____ (4.3). Another important symbol is the one for "the sum of," (14)_____ (4.3). The symbol (15)_____ (4.3) represents any unspecified observation. The mean serves as the balance point for a distribution. This is true because the sum of all observations, expressed as positive and negative deviations from the mean, always equals (16)_____ (4.3).

Similar values of the mode, median, and mean will usually indicate that the distribution is (17)_____ (4.4). In this case, (18)_____ (4.4) can be used to describe the distribution. If a distribution is skewed by extreme observations, the values of the three measures of (19)_____ (4.4) differ considerably. Since the (20)_____ and the (21)_____ (4.4) are not sensitive to extreme observations, the (22)_____ (4.4) should be reported along with the median. Accordingly, large differences between the mean and the median signal a (23)_____ (4.4) distribution. If the mean is larger than the median, the distribution will be (24)_____ (4.4) skewed, but if the median is larger than the mean, the distribution will be (25)_____ (4.4) skewed. Overall, the (26)_____ (4.5) is the most preferred average.

For qualitative data, the (27)_____ (4.5) is always appropriate. However, the (28)_____ (4.5) is only appropriate if the data can be ordered.

Problems and Exercises

Formula for the mean: $\overline{X} = \dfrac{\Sigma X}{n}$

1. A number of college students were surveyed to determine what form of music they most often purchase. For the following results, which type of average is appropriate and why?

   a. cassette tape   135
   b. compact disc   98
   c. 45-rpm record   53
   d. albums   222

2. A university gymnastics team wins "best team" honors at the regional spring meet. The following are the individual awards which contributed toward the team win. Determine which averages, if any, would be appropriate and calculate the appropriate value(s).

1st place	3rd place	5th place	1st place
4th place	3rd place	6th place	3rd place
2nd place	7th place	3rd place	2nd place
3rd place	5th place	3rd place	1st place

3. The sleep research center at a medical school selected volunteers for research. The ages of the volunteers are as follows. Compute the mean, median, and mode.

   22, 45, 34, 25, 23, 36, 23, 38, 27, 32, 44, 55

4. At the county fair the ring toss game cost $1 for ten rings. The first 15 customers achieved the following results. Calculate the mean, median, and mode.

6, 4, 9, 5, 6, 6, 2, 6, 4, 5, 8, 8, 7, 9, 3

5. The following grades were earned by students on the first statistics exam of the term. Calculate the mean, median, and mode.

98, 85, 93, 77, 76, 99, 89, 75, 93, 92, 90, 77, 93, 86

Beyond the Basics

Chapter 4 introduces both the first symbols and the first formula. You will have to deal with many more in subsequent chapters. Here is a hint about memorizing them. Get some 3 x 5 cards. On one side of a card write the symbol or formula. On the other side, write the meaning of the symbol or some pertinent information about the formula. Study them just as you did flash cards when you were younger. The type of rote learning required for this information is often tedious. You will learn best from frequent repetitions of short duration. The 3 x 5 cards are easy to put in your purse or pocket. Keep them handy for frequent review whenever you have a few spare minutes. This type of review will be far more effective than hours of concentrated time. You may recall from a General Psychology class the proven learning principle that distributed practice is more effective than massed practice. Here is an opportunity to put some of that theory to practical use.

Post Test

1. What measure(s) of central tendency is best when extreme observations are present in the data set?

2. What measure of central tendency is appropriate for qualitative data?

3. _____ is the most frequently occurring observation.

4. What is the symbol for the mean? _____

5. What is the formula for the mean? _____

6. What is the symbol for a single observation? _____

7. What does a bimodal distribution suggest about the data set?

8. What measure(s) of central tendency should be reported for a skewed distribution?

9. The single most preferred average is the _____.

10. When the median exceeds the mean, the distribution is _____ skewed.

Answers

<u>Text Review</u>

1.	measures of central tendency	15.	$\overline{X}$
2.	mode	16.	$\overline{0}$
3.	prevalent or fashionable	17.	not too skewed
4.	bimodal	18.	any of them
5.	multimodal	19.	central tendency
6.	differences or subsets	20.	mode
7.	median	21.	median
8.	50	22.	mean
9.	value	23.	skewed
10.	mean	24.	positively
11.	mean	25.	negatively
12.	n	26.	mean
13.	$\overline{X}$ (mean)	27.	mode
14.	Σ (sum of)	28.	median

<u>Problems</u>

1. Only the mode is appropriate, as this is qualitative data that cannot be ranked.
2. Since this set of qualitative data can be ranked from most desirable, or highest prize (1st place) to least desirable, or lowest prize (7th place), both the mode and median should be reported.

1st place	-	3	mode = 3rd place
2nd place	-	2	median = 3rd place
3rd place	-	6	
4th place	-	1	
5th place	-	2	
6th place	-	1	
7th place	-	1	

3. mean = 33.67   mode = 23   median = 33
4. mean = 5.87   mode = 6   median = 6
5. mean = 87.36   mode = 93   median = 89.5

<u>Post Test</u>

1. mean and median
2. mode always, median if the data can be ranked
3. the mode
4. the symbol for the mean is $\overline{X}$
5. the formula for the mean is $\dfrac{\Sigma x}{n}$
6. the symbol for a single observation is x
7. a bimodal distribution indicates differences or subsets among the observations in the distribution
8. both the mean and the median
9. mean
10. negatively skewed

DESCRIBING VARIABILITY

Learning Objectives

You should develop knowledge and understanding of the key terms.

You will be able to calculate and interpret the various measures of variability.

Key Terms

Measures of variability -- General term for various measures of the amount of variation or differences among observations in a distribution.

Range -- The diffference between the largest and smallest observations.

Variance ($\underline{S}^2$) -- The mean of the squared deviations.

Interquartile Range (IQR) -- The range for the middle 50 percent of all observations.

Standard Deviation (S) -- The square root of the mean of the squared deviations.

Text Review

An exact measure of variability representing the difference between the largest and smallest observation is the (1) _Range_ (5.2). The range has the advantage of being easily understood and easily (2) _Calculated_ (5.2). However, it also has shortcomings. One weakness is that it is based on (3) _2 observations_ (5.2). In addition, the value of the range tends to (4) _increase_ (5.2) as the number of observations increases.

An important spinoff of the range is the (5) _IQR_ (5.3), which represents the range for the middle (6) _50%_ (5.3) of the distribution. Like the range, the interquartile range is fairly easy to calculate, but unlike the range, it is not sensitive to (7) _Extreme_ (5.3).

Therefore, the interquartile range might be especially appropriate when there are (8) _Outliers_ (5.3) in the distribution.

The (9) _Variance_ and (10) _Standard Vari._ (5.4) are the preferred measures of variability and hold the same exalted position as the (11) _Mean_ (5.4) holds among measures of central tendency. Because the variance is calculated by expressing observations as deviations from the mean and dividing by the number of observations, it qualifies as a type of mean. There are two formulas for calculating the variance, the (12) _definition_ (5.5), which helps to understand the origin of the variance, and the (13) _computation_ (5.6), which is better to use for calculation when the mean is a complex number or the number of observations is large.

Since the unit of measure becomes illogical if squared as in the calculation of the variance, a new measure called (14)_____ (5.7), must be used. The standard deviation is simply the (15)_____ (5.7) of the variance. Between the variance and the standard deviation, the (16)_____ (5.7) is the preferred measure of variability. The standard deviation has the advantage of expressing the observations in (17)_____ (5.8) units of measure.

The standard deviation may be computed from two formulas, the (18)_____ and the (19)_____ (5.8). The (20)_____ formula is the more convenient for calculation. The standard deviation may be thought of as a (21)_____ measure of the (22)_____ amount by which observations deviate from their (23)_____ (5.9).

For most frequency distributions, a majority of observations, as many as (24)_____ percent (5.10), are within one standard deviation of the mean. By the same token, a minority, as few as (25)_____ percent (5.10), will deviate more than two standard deviations from the mean.

The mean is a measure of (26)_____ (5.11), and the standard deviation is a measure of (27)_____ (5.11) from the mean. Standard deviations may be expressed as positive or (28)_____ (5.11) deviations from the mean, but standard deviations cannot be negative numbers. The positive and negative signs indicate direction from the mean.

Measures of variability for qualitative data are virtually (29)_____ (5.12). That is, the range, variance, and standard deviation cannot be computed. However, descriptive terms, such as <u>maximum</u>, <u>minimum</u>, or <u>intermediate</u> variability, can be used to describe qualitative data.

Problems and Exercises

1. Here are the data from Chapter 4, problem 2. Would you describe the variability as maximum, minimum, or intermediate?

1st place	3rd place	5th place	1st place
4th place	3rd place	6th place	3rd place
2nd place	7th place	3rd place	2nd place
3rd place	5th place	3rd place	1st place

2. Using the data from Chapter 4, problem 3, compute the standard deviation and the variance. (Hint: You may use the mean calculated in Chapter 4 and, for easier computation, round off the mean to the nearest whole number $\overline{X}$ = 34) Follow the steps indicated by the table.

X	$(X - \overline{X})$	$(X - \overline{X})^2$
22		
45		
34		
25		
23		
36		
23		
38		
27		
32		
44		
55		

$$\Sigma (X - \overline{X})^2 = $$

Note: If the actual mean of 33.67 had been used, it would be necessary to use the computational formula rather than the definition formula. You might find it interesting to calculate the standard deviation and variance both ways and see what difference is made by rounding off the mean from 33.67 to 34.

3. Round off the mean and the standard deviation in problem 2 to the nearest whole numbers and answer the following questions.
   a. Between what two scores would 68 percent of the observations fall?
   b. Beyond what scores would 5 percent or less of the observations fall?
   c. What score would be one standard deviation above the mean?
   d. Where would a score of 40 fall?
   e. Where would a score of 21 fall?

4. Compute the range for the data in problem 2.

Beyond the Basics

Here is a crossword puzzle to help you review some of the key terms from the first five chapters.

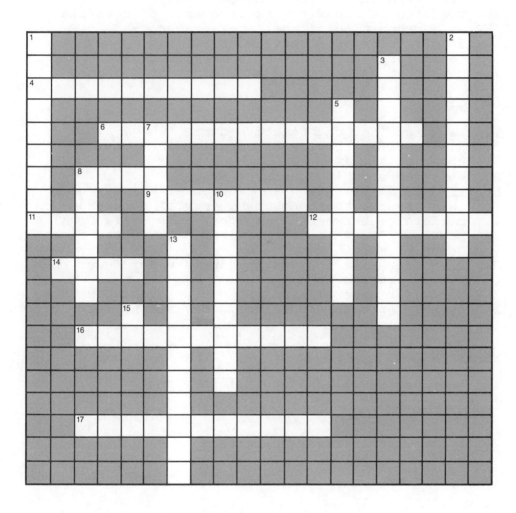

Across Clues

4. study of tools for research
6. percentage of observations with a similar or smaller value
8. sum of observations divided by the number of observations
9. frequency distribution appropriate when more than twenty observations are present
11. most frequently occurring observation
12. distribution showing the frequency of each class as a proportion of the total frequency
14. a collection of observations
16. tools for organizing and summarizing data
17. data that is expressed as a class or category

Down Clues

1. bar graph for quantitative data
2. distribution showing the total number of observations at or below each class
3. data representing count or amount
5. kind of distribution sorting observations into classes that show frequency of occurrence
7. difference between the largest and smallest value
8. middle value when observations are ordered from least to most
10. frequency distribution appropriate for fewer than twenty observations
13. tools for generalizing beyond the available data
15. symbol for mean and standard deviation, respectively

Post Test

1. What are two limitations of the range?

2. Which is the preferred measure of variability? Why?

3. What measures of variability are appropriate for qualitative data?

32

4. Why is the variance referred to as a type of mean?

5. What conditions indicate that the computation formula should be used to calculate standard deviation or variance?

Answers

Text Review

1. range
2. calculated
3. only two observations
4. increase
5. interquartile range
6. 50 percent
7. extreme observations
8. outliers
9. variance
10. standard deviation
11. mean
12. definition
13. computation
14. standard deviation
15. square root
16. standard deviation
17. original
18. definition
19. computation
20. computation
21. rough
22. average
23. mean
24. 68 percent
25. 5
26. position
27. distance
28. negative
29. nonexistent

Problems and Exercises

1. The variability is intermediate.

2.

$\underline{X}$	$(\underline{X} - \underline{\bar{X}})$	$(\underline{X} - \underline{\bar{X}})^2$
22	-12	144
45	11	121
34	0	0
25	-9	81
23	-11	121
36	2	4
23	-11	121
38	4	16
27	-7	49
32	-2	4
44	10	100
55	21	441

$$(\underline{X} - \underline{\bar{X}})^2 = 1202$$

$\underline{S} = 10.96$, round off to 11 for problem 3
$\overline{S} = 10.01$, round off to 10 for problem 3

3. a. 24 and 44
   b. 14 and 54
   c. 45 44 $\underline{\bar{X}} = 34$ (33.667)
   d. within one standard deviation in the positive direction (above the mean)
   e. within two standard deviations in the negative direction (below the mean)

4. the range = 33

Beyond the Basics

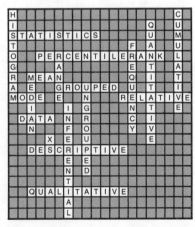

34

1. Two limitations of the range are (a) it increases as the number of observations increases and (b) it is based on only two observations.

2. The standard deviation is the preferred measure of variability because of the previously discussed limitations of the range and the fact that the variance distorts the unit of measure and the standard deviation does not.

3. Measures of variability for qualitative data are virtually nonexistent.

4. The variance is referred to as a type of mean because it is derived from a sum of the raw scores expressed as deviations from their mean, divided by the number of observations. This is the same basic calculation as that for the mean of a data set.

5. When the mean is a complex value or the data set contains a large number of observations, the computation formula is preferred.

Learning Objectives

You should develop knowledge and understanding of key terms.

You will understand the theoretical concepts of the normal curve and the standard normal table.

You will be able to use the normal curve and the standard normal table to solve a variety of problems.

You will be able to calculate and interpret transformed scores such as z scores.

Key Terms

Normal curve -- A theoretical curve noted for its symmetrical bell-shaped form.

z score -- A score that indicates how many standard deviations an observation is from the mean of the distribution.

Standard normal curve -- The one tabled normal curve with a mean of 0 and a standard deviation of 1.

Standard score -- Any score expressed relative to a known mean and standard deviation.

Transformed standard score -- A standard score that, unlike a z score, usually lacks negative signs and decimal points.

Text Review

The normal curve is a (1)_____ (6.2) curve noted for its (2)_____(6.2) bell-shaped form. Because of the symmetrical shape, the lower half is a (3)_____(6.2) image of the upper half. The curve peaks at a point (4)_____ (6.2) above the horizontal spread and then tapers off at each end. Theoretically, these tails of the curve extend (5)_____ (6.2), as they never touch the horizontal axis. In order to use the normal curve, we must assume that a set of quantitative data is (6.2) (6)_____. We must also have the values of the (7)_____ and the (8)_____ (6.2).

Another requirement for using the normal curve is that the original observations must be expressed as (9) _____ (6.3), indicating how many standard deviations an observation is from (10)_____ (6.3). The z score always indicates the value of the original score relative to its (11)_____ and (12)_____ (6.3).

The standard normal curve always has a mean of (13)_____ and a standard deviation of (14)_____ (6.4). A major advantage of using the standard normal curve is that a table is available from which we can determine the proportion represented by area under the curve at any position on the horizontal axis marked by a particular z score. The standard normal table consists of columns of z scores coordinated with columns of (15)_____ (6.5).

In the standard normal table, columns B and B′ represent the proportion of area under the normal curve between the mean and a z score value. Columns C and C′ represent the area between the z score and the tail of the curve. Always remember that the B and C columns in each half will sum to (16)_____ (6.5), and the total area under the normal curve always equals (17)_____ (6.5). When solving problems with the standard normal table, concentrate on the (18)_____ (6.5) and don't try to memorize solutions.

Any score expressed relative to a known mean and standard deviation is referred to as a (19)_____(6.8). Z scores qualify as standard scores, even when the original distribution is not normal and the standard normal table cannot be used. Often, z scores are used to report test results because they give us information about (20)_____ (6.8) performance on one or more tests. Since z scores reflect performance relative to a group rather than an absolute standard, it is important to know the nature of the reference group.

Although z scores are the most important standard scores, it is sometimes desirable to convert them to (21)_____ (6.9) scores, which lack decimals and negative signs. An example of transformed standard scores are (22)_____ (6.9), which have a mean of 50 and a standard deviation of 10. Standard scores are often preferred over percentile ranks when interpreting test scores because they are equally spaced along the horizontal scale. A major weakness of the percentile ranks is that they lack the orderly increase in values apparent in standard scores.

Problems and Exercises

1. Assume that ACT Composite scores approximate a
   normal curve with a mean of 18 and a standard
   deviation of 6. Using the formula for z scores,
   convert the following raw scores to z scores.

   $$z = \frac{X - \overline{X}}{S}$$

   a. 32
   b. 21
   c. 14
   d. 25
   e. 16

2. Using the standard normal table, find the percentile
   rank for each of the preceding scores. (Hint: The
   area under the curve represents proportion, which is
   the same as percent without the decimal.)

3. Finding proportion for a score left of the mean.
   Assume that women's shoe size approximates a normal
   distribution with a mean of 7 and a standard
   deviation of 1.5. What proportion of women will
   wear size 5 or smaller?

Step 1 - Sketch the normal curve and identify the target area.

_____
2.5   4.0   5.5   7   8.5   10.0   11.5

Step 2 - Express the size 5 (target score) as a z score.

$$\frac{5 - 7}{1.5} = \frac{-2}{1.5} = -1.33$$

Step 3 - In column A of the standard normal table, locate the z score of 1.33. From the sketch in step 1, you can see that you are looking for an area at the tail of the curve to the left of the mean. Therefore, you would read the proportion value in C' corresponding to the z score of -1.33. The answer is .0918, or 9%.

Now try one on your own. Using the same data in problem 3, what proportion of women will wear size 4.5 or smaller? Follow the same steps as above. Sketching the normal curve is extremely important, as it helps you with a visual interpretation. Don't be tempted to omit this step.

4. Finding proportion for a score right of the mean.
    Using the data from problem 3, what proportion
    of women wear shoes size 9 or smaller?

    Step 1 - Sketch the normal curve as presented
    earlier.

    Step 2 - Express the size 9 as a z score.

    $$\frac{9 - 7}{1.5} \quad = \quad \frac{2}{1.5} \quad = \quad 1.33$$

    Step 3 - In column A of the standard normal table,
             locate the z score of 1.33. From the
             sketch in step 1, you can see that you
             are looking for the area that includes all
             the lower half of the curve plus the area
             from the mean to the position of the z
             score. Therefore, you must take the value
             from column B that corresponds to the z
             score of 1.33 (.4082) and add it to .5000
             (the proportion value of the lower half of
             the curve). The answer is .9082, or round-
             ing off, 91 percent of women wear shoes
             size 9 or smaller. It is little wonder
             that women with larger feet have trouble
             finding shoes since they represent only 9%
             of the market.

    Now try one on your own. What proportion of women
    wear size 8 or smaller?

5. Finding scores.
   Assume the ACT approximates a normal curve with
   a mean of 18 and a standard deviation of 6. Because
   of overcrowding, a small college wants to use the
   test to select applicants who score in the top 25%.
   What would be the appropriate cutoff score?

   Step 1 - Sketch the normal curve and identify the
           target area.

   |_____|
   6   12   18   24   30

   Step 2 - Find z in the standard normal table.
           Since you are looking for the top 25%,
           you would be looking for .25 in column C.
           Locate this figure and the corresponding
           z score. Note that .25 is not listed
           exactly. Instead, you must choose between
           .2514 and .2483. You should use .2514 as
           it is closer in value to .2500. The
           corresponding z score is .67.

   Step 3 - Convert z to the target score (X).

   $X = \overline{X} + (z)(S)$

   $X = 18 + (.67)(6)$ $= 18 + 4.02$ $= 22.02$

   Rounding off, this gives us the answer. The college
   should set its cutoff score at 22.

   Now try one on your own. Suppose the same college
   wants to select students for an honors program. If
   they want students whose scores are in the top 7%,
   what cutoff score would they use. (Hint: Remember
   that 7% is .07 proportion.)

Beyond the Basics

Ask professors from three of your classes to help you with this problem. You will need to know the mean and standard deviation on the last exam you took in each course, as well as your own score. If the professor has not calculated this information, perhaps the data can be made available without disclosing the identity of other students so that you can calculate the mean and standard deviation yourself. Using this information, compute your own z score for each course. Determine which class represents your best performance relative to your classmates.

Post Test

1. The normal curve is a theoretical curve noted for being _____.

2. In order to solve problems using the normal curve, one must have the values of the _____ and the _____.

3. A _____ indicates how many standard deviation units an observation is from its mean.

4. The one tabled normal curve with a mean of 0 and a standard deviation of 1 is the _____.

5. In the standard normal table, _____ would be used to determine the proportion of area under the curve between the mean and a particular z score.

6. The total area under the curve from the mean to the tail on the left (the lower half) represents _____ proportion of the whole curve.

7. The total area under the curve always equals _____.

8. Any score expressed relative to a knowm mean and standard deviation is referred to as a _____.

9. A standard score that lacks decimals and negative signs is a _____.

10. The _____ indicates the percentage of scores in the distribution with similar or smaller values.

Answers

Text Review

1. theoretical
2. symmetrical
3. mirror
4. midway
5. infinitely
6. normally distributed
7. mean
8. standard deviation
9. z scores
10. the mean
11. mean
12. standard deviation
13. zero
14. one
15. proportions
16. .5000
17. 1.000
18. logic
19. standard score
20. relative
21. transformed standard
22. T scores

Problems and Exercises

1. a. 2.33   b. .5   c. -.67   d. 1.17   e. -.33

2. a. 99   b. 69   c. 25   d. 88   e. 37

3. $z = -1.67$, under C' column area = .0475 = 5%

4. $z = .66$, under B column area = .2454, add to .5000 for lower half of curve = .7454 = 75%

5. When $C = .07$, $z = 1.48$ because .0694 is closer to .07 than .0708 is, $X = 26.88$ or 27. Note: It is important to round off to 27, as ACT scores are reported in whole numbers. Keep in mind the original unit of measure when reporting values of X.

Post Test

1. being symmetrical, bell-shaped
2. mean and standard deviation
3. z score
4. standard normal curve
5. column B
6. .5000
7. 1.000
8. standard score
9. transformed standard score
10. percentile rank

Learning Objectives

You should develop knowledge and understanding of key terms.

You will be able to calculate and interpret Pearson correlation coefficients.

You will be able to interpret scatterplots.

Key Terms

Positive relationship -- When pairs of observations tend to occupy similar relative positions in their respective distributions.

Negative relationship -- When pairs of observations tend to occupy dissimilar and opposite positions in their respective distributions.

Scatterplot -- A special graph containing a cluster of dots that represents all pairs of observations.

Pearson correlation coefficient -- A number between -1 and +1 that describes the linear relationship between pairs of quantitative variables.

Linear relationship -- A relationship that can be described with a straight line.

Curvilinear relationship -- A relationship that can be described with a curved line.

Correlation coefficient -- A number between -1 and 1 that describes the relationship between pairs of variables.

Correlation matrix -- Table showing correlations for all possible pairs of variables.

Text Review

In previous chapters, we have examined individual data sets representing collections of records or observations of some characteristic that varied among individuals, (e.g., height, weight, and IQ score for people, or popping time for kernels of corn and average burning time for light bulbs). In statistics, since the values of these characteristics vary among individauls, they are commonly referred to as variables. In Chapter 7 we will examine relationships between two variables. Therefore, we will see pairs of observations.

When relatively high values of one variable are paired with relatively high values of the other variable, and low values are paired with low values, the relationship is (1)_____ (7.1).

Another way to think of this is that values of one variable increase as values of the other increase. Likewise, values of one variable decrease as values of the other decrease. An example of a positive relationship between two variables would be study time and performance in statistics class. As study time increases, performance in class will also increase. Thus, relatively high values of each variable are paired and relatively low values of each are paired.

When pairs of observations occupy dissimilar and opposite relative positions in their respective distributions, the relationship is (2)_____ (7.1). An example of a negative relationship would be auto gas mileage and horsepower. As horsepower is increased, gas mileage would be expected to decrease. Conversely, when horsepower is decreased, gas mileage should increase.

It may occur to you that certain variables may exist which would not be related either positively or negatively. This happens to be true. Consider, for example, hat size and IQ. Pairs of these variables would not occupy either similar or dissimilar positions in their respective distributions. If the pairs were graphed on a scatterplot, no pattern would appear. These variables would be said to have no relationship. A calculated correlation coefficient for the two variables would be near zero.

The graph which shows relationship between variables as a cluster of dots is called a (3)_____ (7.2). The pattern formed by the dot cluster is significant. If the cluster has a slope from upper right to lower left, it depicts a (4)_____ (7.2) relationship. If the slope is from upper left to lower right, the

45

relationship is (5)_____ (7.2). A dot cluster that lacks any apparent slope reflects (6)_____ (7.2). The more closely a dot cluster approximates a straight line, the (7)_____ (7.2) the relationship. When a relationship can be described with a straight line, it is called a (8)_____ (7.2) relationship. When the dot cluster forms a curved line, the relationship is said to be (9)_____(7.2).

The relationship between two variables that represent quantitative data is described by a correlation coefficient and designated by the symbol (10)_____ (7.3). The correlation coefficient ranges in value from (11)_____ to (12)_____ (7.3). The sign of r indicates whether the relationship is (13)_____ or (14)_____ (7.3) The value of r indicates the (15)_____ (7.3) of the relationship. The correlation coefficient is referred to as (16)_____ (7.3) and was named after the British scientist Karl Pearson.

Interpretation of r is related to the direction and strength of the correlation. The direction, either (17)_____ or (18)_____ (7.6), is indicated by the sign of the correlation coefficient. The strength is reflected by the (19)_____ (7.6) of r. An r value of .50 or more in either direction is typical of important relationships in most areas of behavioral and educational research. For test reliability studies, however, a value of .80 is the minimum acceptable value.

In order to understand r^2 (the correlation coefficient squared) think for a moment about variability in a single distribution, which we studied in Chapter 5. In a single variable, we described variability with the measure of the standard deviation. We studied the variability graphically by looking at the shape of the frequency polygon. In a scatter plot, two variables are depicted graphically. Imagine a frequency polygon along the horizontal axis of the scatter plot. This shows the shape of the distribution for variable X. Imagine also a frequency polygon along the vertical axis of the scatter plot. This shows the shape of the distribution for variable Y. As we examine the relationship between the two variables, we know that some of the variability in Y must be due to the variability in X.

Let's substitute real data and think this through. Variable X represents SAT scores. Variable Y represents college GPA. These variables constitute a strong positive relationship, approximately .57. There must be many reasons why GPA in college would vary. Some of that variability is probably due to preparation for college as reflected by SAT scores. The remaining variability might be due to such factors as whether the student works, how the adjustment is made to living away from home, and how many hours per week are devoted to studying or partying. The actual amount of variability in college GPA, variable Y, that can be explained by the variability in X, SAT scores is reflected by the value of r^2. Thus we compute .57 squared = .3249, or .32. We interpret this value by saying that 32 percent of the variability of Y can be explained by the variability in X. The remaining 68 percent of the variability of Y would be explained by a combination of many other factors or variables, probably some of those mentioned earlier.

One important concept to keep in mind is that a correlation coefficient never provides information about cause and effect. Cause and effect can only be proved by (20)_____(7.8).

Problems and Exercises

Some hints for computing and interpreting correlation coefficients.

a. Since a correlation coefficient always has a value between -1 and +1, you will know you have made an error if your answer is outside this range.

b. It may help to sketch a scatter plot before you start computing. This would help you to estimate whether the r value will be positive or negative and how strong the relationship may be.

c. When computing r, remember that n represents number of pairs of X and Y values.

1. A physical education teacher suspects that eye-hand
    coordination is an important factor in performance
    in both shooting a basketball and pitching horse-
    shoes. Eight student volunteers achieved the
    following results out of fifteen tries each at
    shooting the basketball and pitching horseshoes.
    Compute a Pearson r and answer the questions which
    follow.

Basketball X	Horseshoes Y
8	2
10	3
7	3
6	5
7	6
4	8
2	9
3	10

a. In reality, eye-hand coordination probably is
    related to performance in both sports. What
    possible explanation can you offer for the negative
    correlation obtained in this particular instance?

b. What proportion of the variability of horsehoe
    performance is explained by variability in
    basketball performance?

2. Sketch a scatter plot that depicts the relationship
    for problem 1.

3. A statistics professor asked her students to record the number of hours spent studying for an exam. She then computed r for the correlation between hours spent studying and exam grades.
   a. What kind of r value would she have been likely to find?
   b. Assume the computation resulted in an r value of +.63. Write a verbal statement interpreting the results.
   c. What other variables might account for the variability in the exam scores which cannot be attributed to study time.

4. A study was conducted to determine the relationship between age and strength of handgrip. (Handgrip is measure by pounds of pressure placed on a mechanical device.)
   a. Sketch a scatterplot.
   b. Calculate the r value for the following data.
   c. Comment on the shape of the scatterplot.

Age	Handgrip
4	5
7	8
13	11
18	22
29	24
38	24
47	23
58	21
67	17
76	13

5. The director of institutional research at a small college wants to confirm the relationship between SAT scores and GPA. Using a small sample, he obtains the following results. Use the z score formula to calculate r. Interpret your results.

SAT	GPA
z_X	z_Y
.5	.4
1.2	1.0
.7	.4
2.3	1.7
-1.0	-.7
-.8	-.5
1.6	1.3
-1.4	1.8

Beyond the Basics

Make an appointment to visit the counseling/testing center at your school. Ask if someone will review with you the examiner's manuals of any achievement, aptitude, or personality tests that they may have. Look for the section on test reliability and validity. Are there any studies reported in which correlation has been used to demonstrate reliability? What are the variables that were correlated? Is the correlation coefficient high enough to meet the standard set in Chapter 7 for test reliability, r = .80 or higher?

Post Test

1. What is the possible range of values for a correlation coefficient?

2. What is the meaning of an r value near 0?

3. Comment on the use of correlation to show cause and effect.

4. What is the meaning of an r value of +1.

5. What is the meaning of an r^2 value of .36?

6. The results of correlations for all possible pairs of variables in a given study can be shown in a _____.

7. A relationship described with a curved line is called _____.

Answers

Text Review

1. positive
2. negative
3. scatterplot
4. positive
5. negative
6. no relationship
7. stronger
8. linear
9. curvilinear
10. r

11. -1
12. +1
13. positive
14. negative
15. strength
16. Pearson r
17. positive
18. negative
19. value
20. controlled laboratory experiment

Problems and Exercises

1. a. Effects of chance, practice, motivation, fatigue, other factors are also possible; $r = -.90$
    b. $r^2 = .81$, or 81%

2. The scatterplot is approximately linear, sloping from upper left to lower right.

3. a. Reasonably strong positive
    b. There appears to be a strong positive relationship between study time and performance on statistics exams.
    c. Test anxiety, health, fatigue, motivation, and other worries or stress, to name a few

4. b. $r = .398$, or .40
    c. The scatterplot is an inverted U.

5. $r = .78$.   There is a strong positive relationship.

Post Test

1. -1 to +1
2. No relationship between the two variables
3. Correlation can never show cause and effect
4. Perfect positive relationship
5. Thirty-six percent of the variability of Y is explained by the variability of X.
6. correlation matrix
7. curvilinear

Learning Objectives

You should develop knowledge and understanding of the key terms.

You will be able to calculate and interpret prediction intervals.

Key Terms

Least squares prediction equation -- The equation that minimizes the total of all squared prediction errors for known $\underline{Y}$ scores in the original correlation analysis.

Prediction interval -- A range of values that covers the unknown $\underline{Y}$ score with a known degree of confidence.

Standard error of prediction -- A rough measure of the average amount by which known $\underline{Y}$ values deviate from their predicted $\underline{Y}'$ values.

Variance interpretation of $\underline{r}$ -- The proportion of variance explained by, or predictable from, the existing correlation.

Text Review

By "predicting" what is known in a correlation of two variables, we can predict what is (1)_____ (8.2). This is accomplished by placing a (2)_____ (8.2) line in such a way that it passes through the main cluster of dots in the scatterplot. Positive and negative errors are avoided by squaring the difference between the predicted value and the actual value. Thus, the prediction line is referred to as the (3)_____ _____ or the (4)_____ (8.3).

The search for the least squares prediction line would be a frustrating trial and error process, except for the precision of the least squares equation, (5)_____ (8.4). In this equation, $\underline{Y}'$ represents the (6)_____ (8.4) value, and $\underline{X}$ represents the (7)_____ (8.4) value. The other values in the equation must be computed. When the computation is complete, the equation has the very desirable property of minimizing the total of all squared predictive errors for known values of Y in the original correlation analysis. Note in the Key Terms section earlier that this essentially defines the least squares prediction equation.

Two limitations exist for the application of the least squares prediction equation. One is that predictions may not be reliable if extended beyond the maximum value of X in the original data. Second, since there is no proof of cause-effect in correlation, the desired effect simply may not occur.

Graphs may be constructed to depict the prediction equation. However, this should be done for (8)_____ (8.5) purposes and not for prediction. It is more accurate to make the actual predictions from the (9)_____ (8.5)

The least squares prediction equation is designed to reduce error in prediction, but it does not eliminate it. Therefore, we must estimate the amount of error, understanding that the smaller the error, the more accurate our prediction. The estimated predictive error is expressed by the (10)_____ (8.6). This represents a rough measure of the average amount by which known Y values deviate from their predicted Y' values.

The value of r is extremely important in relation to predictive error. When r = 1, the predictive error will be (11)_____ (8.6). The most accurate predictions can be made when r values represent (12)_____ (8.6) relationships, whether positive or negative. Prediction should not be attempted when r values are (13)_____ (8.6), representing weak or nonexisting relationships.

We can increase our accuracy of prediction by predicting a range of possible Y values instead of a single value. We can even be 95% confident that our predicted range of values includes the actual value. Our claim of 95% confidence indicates that approximately 95% of the predicted Y' values for anybody who could conceivably have an observed value of this variable would be within the range we predicted. How do we establish the range of values called (14)_____ (8.9). Assume that predictive errors approximate a normal distribution. Then the predicted Y' value becomes the mean of the distribution and the standard deviation is equal to the

54

standard error of prediction. From the standard normal table (Chapter 6) we recall that the z values that define the middle 95% of the total area under the curve are +1.96. Thus, 95% of the predicted values should lie between values obtained by adding and subtracting 1.96 standard error of prediction units to the predicted value of Y'. Remember, new intervals must be constructed for any new prediction based on a different value of X. Also, new levels of confidence can be specified by changing to an appropriate value of z.

There are some assumptions which must be met in order to apply the concepts of prediction we have been discussing. One is that using the prediction equation requires the underlying relationship to be (15)_____ (8.10). Therefore, if the scatterplot for an original correlation analysis is (16)_____ (8.10), this procedure would not be appropriate. A second assumption is that the dots in the original scatterplot will be dispersed equally about all segments of the prediction line. This is known as (17)_____ (8.10). The third assumption is that for any given value of X, the corresponding distribution of Y values is (18)_____ (8.10) distributed. The final assumption is that the original data set of paired observations must be fairly large, usually in the hundreds.

Problems and Exercises

1. Recall the correlation problem from Chapter 7, examining the relationship between shooting a basketball (X) and pitching horseshoes (Y). Using a new set of data, a positive correlation of .85 was found. The means and standard deviations are as follows:

$\overline{X}$ = 8.75 $\overline{Y}$ = 7.75

$\underline{S}_x$ = 2.12 $\underline{S}_y$ = 2.25

a. Determine the least squares equation for predicting number of ringers in horshoes from number of baskets made in basketball.

b. Calculate the standard error of prediction, $\underline{S}_{y \cdot x}$.

c. Predict the number of horseshoe ringers made by Larry who made eleven baskets out of the fifteen attempts.

d. Construct a 99% prediction interval for Larry's horshoe pitching score.

e. Interpret the prediction interval.

f. Predict the horshoe pitching score for Earl who scored only five baskets.

g. Construct the 95% prediction interval for this score.

h. Interpret the prediction interval.

56

2. For more practice, return to Chapter 7 and change the values of Y to create a lower positive correlation between horseshoe pitching and basketball. Then answer the preceding questions. Don't forget, you must calculate a new mean and standard deviation for Y, as well as the new correlation coefficient. The values for X will remain the same.

Beyond the Basics

Discuss with one of your classmates some practical applications for prediction techniques in applied situations in such fields as education, business, medicine, and politics. You might also locate professional journals from the one area that interests you most and try to find reports of research using prediction techniques. You will be pleased at your new ability to understand the more statistical aspects as you read.

Post Test

Answer each of the following questions true or false.

1. When the value of r equals either + 1.00 or -1.00, the standard error must equal 0.

2. In a normal bivariate distribution, the mean of Y will equal the mean of X.

3. When the value of r increases, the standard error of the estimate decreases.

4. In a scatterplot, the predicted variable is usually on the Y axis.

5. The standard error of estimate is a type of standard deviation.

6. The plotting of a regression line is not logical unless the value of r differs significantly from 0.

7. The least squares prediction equation minimizes the total of all squared predictive errors for known Y scores in the original correlation analysis.

8. The value of $1 - r^2$ indicates the proportion of variability in $\underline{Y}$ that is predicted by $\underline{X}$.

9. The value of $\underline{r}^2$ can be interpreted as the proportion of explained variance.

10. Individual scores can be described by $\underline{r}^2$.

Answers

Text Review

1.	unknown	11.	zero
2.	prediction	12.	strong
3.	least squares regression line	13.	low
4.	least squares prediction line	14.	prediction intervals
5.	$\underline{Y}' = b\underline{X} + \underline{a}$	15.	linear
6.	predicted	16.	curvilinear
7.	known	17.	homoscedasticity
8.	descriptive	18.	normally
9.	least squares prediction equation		
10.	standard error of prediction		

Problems and Exercises

1. a. $\underline{Y}' = (b)(\underline{X}) + \underline{a}$
   $= (.90)(\underline{X}) + -.13$

   b. 1.19

   c. $\underline{Y}' = (.90)(11) + -.13 = 9.77$, or 10 ringers

   d. $10 \pm (2.58)(1.19) = 10 \pm (3.07) = 6.93$ to 13.07

   e. It can be predicted with 99 percent confidence that the interval between 6.93 and 13.07 ringers describes Larry's horseshoe performance.

   f. $\underline{Y}' = (.90)(5) + -.13$
   $\underline{Y}' = 4.50 -.13 = 4.37$, or 4 ringers

   g. $4 \pm (1.96)(1.19) = 2.33 = 1.67$ to 6.33

h. It can be predicted with 95 percent confidence
    that the interval between 1.67 and 6.3 ringers
    describes Earl's horseshoe pitching performance.

Post Test

1. true
2. false
3. true
4. true
5. true
6. true
7. true
8. false
9. true
10. false

Learning Objectives

You should develop knowledge and understanding of key terms.

You will be able to calculate and interpret probability problems.

You will be able to use the table of random numbers and the "fishbowl" method to select a random sample from a population.

Key Terms

Population -- Any set of potential observations.

Sample -- Any subset of actual observations from a population.

Random sample -- A sample produced when all potential observations in the population have equal chances of being selected.

Independent outcomes -- The occurrence of one outcome has no effect on the probability that the other outcome will occur.

Conditional probability -- Probability of one event, given the occurrence of another event.

Probability -- The proportion or fraction of times that a particular outcome will occur.

Addition rule -- Add together the separate probabilities of several mutually exclusive outcomes to find the probability that any one of these outcomes will occur.

Multiplication rule -- Multiply together the separate probabilities of several independent outcomes to find the probability that these outcomes will occur together.

Mutually exculsive outcomes -- Outcomes that can't occur together.

Text Review

Populations may be (1)_____ or (2)_____ (9.1).
A real population is one in which all observations are
(3)_____ (9.1) at the time of sampling. A
hypothetical population is one in which all observations
are (4)_____ (9.1) at the time of (9.1)
(5)_____. Often it is not convenient or even
possible to include all observations in a research
project. In such cases, a (6)_____ (9.3) or
subset of observations is taken. The size of the sample
is partially determined by estimated (7)_____
(9.3) among observations and by an acceptable amount of
(8)_____ (9.3)

In order to use inferential statistics, the analysis must
be based on a (9)_____ (9.4) sample. A sample is
random if, at each stage of the sampling, the selection
process guarantees that all remaining (10)_____ (9.4)
have (11)_____ (9.4) chances of being selected.
The observations in a randomly selected sample should be
(12)_____ (9.4) of those in the population.
However, there is no guarantee of this. The term <u>random</u>
describes the process, and not necessarily the outcome.

One of the best-known techniques for selecting a random
sample is the (13)_____ (9.4) method. All
observations must be represented on slips of paper that
are deposited in a bowl and (14)_____ (9.4). The
thorough stirring is a very important aspect of this
method of sample selection.

Another method for generating a random sample involves
the use of the table of (15)_____ (9.5) numbers. When
using this table, the number of digits actually used is
determined by the (16)_____ (9.5). This
method is not very efficient for obtaining a sample from
a (17)_____ (9.5) population.

In an experiment, although subjects may not be selected
randomly, they should be randomly assigned to either the
experimental or control condition. The purpose of random
assignment is to make sure that, except for (18)_____
(9.7) differences, groups of subjects are similar with
respect to any (19)_____ (9.7).

(20)_____ (9.9) refers to the proportion or
fraction of times that a particular outcome will occur.

Eventually, we will want to specify the probability of a particular outcome. Probability can be determined by (21)_____ or (22)_____ (9.9). Probability values can vary between (23)_____ and (24)_____ (9.9), and a set of probability values will always sum to (25)_____ (9.9). Probabilties for various outcomes can be determined by the (26)_____ and (27)_____ rules (9.9). The addition of probabilities is appropriate when none of the outcomes can occur together. These are called (28)_____ (9.10) outomes. They are usually connected by the word or and require the use of the addition rule. When outcomes are not mutually exclusive, don't forget to adjust the addition rule. When probability questions involve clusters of simple outcomes connected by the word and, use the (29)_____ (9.11) rule. This rule is appropriate because the occurrence of one outcome has no effect on the probability of the other outcome. Outcomes of this kind are called (30)_____ (9.11). When the occurrence of one outcome does affect the probability of the other outcome, the probability of the second outcome must be adjusted to reflect this. The probability of one event, given the occurrence of another event, is referred to as (31)_____ (9.11) probability.

Areas under the theoretical normal curve can be interpreted as (32)_____ (9.12).

Problems and Exercises

1. During the 1987 AFC championship games, the televising network conducted a public opinion poll to determine the best NFL quarterback. Numbers were given so that viewers could call a particular number to register their vote for John Elway, Dan Marino, Doug Williams, or Jim Kelly.

    a. Did this survey employ random sampling techniques?

    b. Comment on how the survey could have been improved.

2. Anagrams are formed by changing the order of letters in a word to make a new word. Listed below are the possible ways the letters of the word TEA can be arranged.

tea   tae   ate   aet   eat   eta

a. If an anagram is chosen at random, what is the probability that it will be an English word, as opposed to a nonsense syllable?

b. What is the probability that a randomly chosen letter arrangement will be either a word or a nonsense syllable?

c. What probability rule did you use to solve part b?

d. What is the probability of randomly choosing the word eat?

e. What is the probability of randomly choosing either the word eat or the word ate?

3. One way to randomly assign subjects in an experiment to either the experimental group or control group is to use a die. When the die is rolled, if an odd number appears, the subject is assigned to the experimental group, and if an even number is rolled, the subject is assigned to the control group.

   a. What should be done to ensure that the groups are equal in number?

   b. What would be another way of using the die to make the assignment of subjects to the two groups?

4. A student takes a statistics exam that consists of multiple-choice items, each with four answer choices. She feels confident of her answers on all but four questions.

   a. What is the probability that she will guess correctly on all four that she doesn't know?

   b. What is the probability that she will get all four of the questions wrong?

   c. What could the instructor do, if anything, to decrease the probability of the student guessing correct answers?

5. Why do multiple-choice tests determine better than true-false tests which students really have knowledge of course material?

6. A man has socks in a drawer mixed at the ratio of 4 pairs of white ones to 5 pairs of black ones. How many single socks would he have to take out without knowing the color before he had a matching pair?

Beyond the Basics

If you have ever been to Las Vegas or gambling casinos elsewhere, think about the odds of the probability of winning in the various games of chance. Which ones, if any, would most favor the patron? Which would be most representative of the theory of probability? Is there anything a would-be gambler could do (short of cheating of course) to be better prepared? You may want to find a book on gambling that would help you answer some of these questions.

Post Test

1. Describe the fish bowl method of selecting a random sample.

2. Define a random sample.

3. What is the purpose of random assignment in an experiment?

4. What is the possible range of values of probabilities?

5. What is the sum of all probabilities in a set?

6. What rule should apply for probability problems when outcomes are mutually exclusive?

7. When should the multiplication rule be used?

8. What is the significance of the fact that probabilities represent area under the curve?

9. Differentiate between a real and a hypothetical population.

Answers

<u>Text Review</u>

1.	real	17.	large
2.	hypothetical	18.	random
3.	accessible	19.	uncontrolled variables
4.	not accessible	20.	probability
5.	sampling	21.	speculation
6.	sample	22.	observation
7.	variability	23.	zero
8.	error	24.	one
9.	random	25.	one
10.	observations	26.	addition
11.	equal	27.	multiplication
12.	representative	28.	mutually exclusive
13.	fish bowl	29.	multiplication
14.	stirred	30.	independent
15.	random	31.	conditional
16.	population size	32.	probabilities

<u>Problems</u>

1. a. No. b. The calls should have been made by the network after randomly selecting the people to be called. People who call in voluntarily are self-selected, not randomly selected. Furthermore, a bias was created by having the calls made while one of the quarterbacks was actually on TV during a game. Logically, his fans would be more likely to be watching that particular game.

2. a. .50   b. 1.00   c. addition
   d. 1/6, or .17   e. 2/6 or .33

3. a. Assign them in pairs. If the first subject is
        assigned to the control group, then the second
        automatically goes to the experimental group.

    b. Numbers 1, 2, and 3 could be assigned to the
        control group and numbers 4, 5, and 6 could be
        assigned to the experimental group.

4. a. .25 x .25 x .25 x .25 = .0039
    b. .75 x .75 x .75 x .75 = .32
    c. Use questions with five answer choices.

5. Because the probability of a correct guess on a
    true-false question is .50, but on a four-option
    multiple-choice question, the probability of a
    correct guess is .25.

6. 3, No matter what color is selected first or
    second, the third one will match one of the other
    two. This is a logic problem that requires no
    calculation. However, it should give you practice
    in seeing what's important and what's not. In this
    case, the ratio of 4 to 5 has nothing to do with
    solving the problem.

Post Test

1. All observations are represented on slips of paper.
    The slips are placed in a container and stirred
    thoroughly. Slips are drawn one at a time until the
    desired sample size is reached.

2. A sample is random if all the observations in the
    population have an equal chance of being selected.

3. Random assignment is done to ensure that groups of
    subjects are similar with respect to any
    uncontrolled variables.

4. Probabilities can range in value from 0 to 1.

5. All probabilities in a set sum to 1.

6. With mutually exclusive outcomes, use the addition
    rule.

7. The multiplication rule should be used with independent outcomes.

8. Since probabilites represent the area under the curve, we can use the standard normal table (z table) to determine exact positions along the horizontal base line corresponding to the area or probability. We can then determine the probability that a particular value will fall within the range of specific z scores. For instance, the standard normal table indicates that .95 is the proportion of the area under the curve that falls between z scores of -1.96 and +1.96.

9. A real population is one in which all observations are accessible at the time of sampling. A hypothetical population is one in which all observations are not accessible.

SAMPLING DISTRIBUTIONS

Learning Objectives

You should develop knowledge and understanding of the key terms.

You will understand the theoretical concepts of the sampling distribution of the mean.

Key Terms

Sampling distribution of the mean -- Distribution of sample means for all possible random samples of a given size from some population.

Standard error of the mean -- A rough measure of the average amount by which sample means deviate from the mean of the sampling distribution (or the population mean).

Central limit theorem -- A statement that the shape of the sampling distribution of the mean will approximate a normal curve if the sample size is sufficiently large.

Text Review

A sampling distribution of means refers to the distribution that would exist if all possible samples of a given size were taken from some population and the mean of each of these samples was calculated. Then a distribution would be created from all these means, just as distributions have been created for sets of observations. A mean could be calculated for this distribution, and various observed sample means could be examined in light of distance from the mean of the sampling distribution. Thus, the sampling distribution allows us to determine whether a particular observed sample mean could be viewed as a common or rare outcome. If the observed sample mean is near the mean of the sampling distribution, then it would be a (1)_____ (10.1) outcome. (Remember the characteristics of the standard normal curve from Chapter 6; 68 percent of the observed sample means would be within one standard deviation of the mean of the sampling distribution.) If the observed sample mean is very different from the mean of the sampling distribution, then it would be viewed as a (2)_____ (10.1) outcome.

In reality, a sampling distribution is not constructed, as even with a computer, it would be a horrendous task. Instead, statistical theory supplies the information we need to understand the important idea of the sampling distribution. A highly simplified example of a sampling distribution is created in Section 10.2 of the text. You may wish to reread this section, carefully studying the accompanying figures and tables. From this example, we learn that all values of the sample mean do not occur with equal probability.

There are new symbols to learn that identify the mean and standard deviation of the sampling distribution and the population.   The Greek letters μ (mu) and σ (sigma) represent the (3)_____ and (4)_____ (10.3) of any population. The Greek letters $\mu_{\bar{x}}$ (mu, sub x, bar) and $\sigma_{\bar{x}}$ (sigma, sub x, bar) represent the (5)_____ and the (6)_____ (10.3) of all sample means in a sampling distribution.   To avoid confusion, the sigma, sub x, bar term is usually referred to as the (7)_____ (10.3).

The sampling distribution of means has a mean, and this mean will always equal the (8)_____ (10.4) mean. Therefore, the two terms may be used interchangeably in inferential statistics so that any claims made about the population mean apply to the mean of the sampling distribution and vice versa. Even when a distinction is made for the purpose of emphasis, it is important to note that the numerical value of the two is always the same.

The sampling distribution also has a standard deviation which is referred to as the (9)_____ (10.5). The standard error of the mean equals the (10)_____ (10.5) of the population divided by the square root of the (11)_____ (10.5). The standard error of the mean is a measure of (12)_____ (10.5) in the sampling distribution. This variability decreases as sample size increases. Therefore, more precise generalizations may be made from samples to populations when sample size is (13)_____ (10.5).

The central limit theorem states that the shape of the sampling distribution of the mean will approximate a normal curve if the sample size is sufficiently large. A sample size between (14)_____ and _____ (10.6) is considered sufficiently large. The fact that the sampling distribution approximates the shape of a normal curve is important because it allows us to make statements about the sampling distribution by referring to the table for the standard normal curve.

Sampling distributions can be constructed for medians, proportions, standard deviations, and variances, as well as for other measures. Therefore, it is necessary to provide a full description and refer to the "sampling distribution of the mean" and never just the "sampling distribution."

Problems and Exercises

True or False

1. The standard deviation of a sampling distribution is the standard error of that distribution.

2. The population mean and the mean of the sampling distribution represent the same theoretical concept, but not the same numerical value.

3. As the sample size decreases, the standard error also decreases.

4. The sampling distribution of the mean is a theoretical concept and would not actually be constructed.

5. According to the central limit theorem, in order for the researcher to make the assumption of normality, the sample size must be between 25 and 100.

6. If a population distribution is skewed, a sampling distribution based on sample size of 50 will also be skewed.

7. The symbol μ represents the population mean.

8. The distribution for the population and the sampling distribution of the mean have in common the fact that one deals with <u>all possible</u> observations and the other with <u>all possible</u> random samples.

9. A sample differs, in that it contains only a subset of the possible observations.

10. Any claims that can be made about the population mean can also be made about the sample mean.

11. Any claims about the population mean can also be made about the mean of the sampling distribution.

12. Random samples do not usually represent the underlying population exactly.

Beyond the Basics

Many times students are tempted to skim lightly or ignore charts, graphs, and tables. In this chapter, these helpful visual aids can be a major factor in understanding the theoretical concept of the sampling distribution of the mean. Go back to your text and carefully study all the visual aids in Chapter 10. A true test of your understanding will be to attempt to explain some of the tables or figures to someone else. If you have not already organized a study group for this class, this would be a good time to do so. Keep the group small; three or four is ideal. Also, try to identify students who are about your same level of ability or understanding or higher. A study group composed entirely of people who are struggling in the course will not be very helpful.

Post Test

1. Match the following symbols with the appropriate measure.

    a. population mean $\mu_{\bar{x}}$
    b. population standard deviation μ
    c. mean of the sampling distribution $\sigma_{\bar{x}}$
    d. standard error of the mean $\bar{x}$
    e. sample mean σ

2. What is the relationship between sample size and variability in the sampling distribution?

3. Define the standard error of the mean.

4. What is the sample size needed to meet the central limit theorem?

5. Explain the relationship between the population mean and the mean of the sampling distribution.

Answers

Text Review

1. common
2. rare
3. mean
4. standard deviation
5. mean
6. standard deviation
7. standard error of the mean

8. population
9. standard error of the mean
10. standard deviation
11. sample size
12. variability
13. larger
14. 25 to 100

Problems and Exercises

1.	true	7.	true
2.	false	8.	true
3.	false	9.	true
4.	true	10.	false
5.	true	11.	true
6.	false	12.	true

Post Test

1. a. population mean μ
   b. population standard deviation σ
   c. mean of the sampling distribution $\mu_{\bar{x}}$
   d. standard error of the mean $\sigma_{\bar{x}}$
   e. sample mean $\bar{X}$

2. As sample size increases, variability decreases.

3. Standard error of the mean is a rough measure of the average amount by which sample means deviate from the mean of the sampling distribution (or the population mean).

4. 25 to 100

5. They are always equal in value.

INTRODUCTION TO HYPOTHESIS TESTING

Learning Objectives

You should develop knowledge and understanding of the key terms.

You will be able to calculate and interpret the z test for a population mean.

You will be able to determine appropriate level of significance for hypothesis tests, write alternate and null hypotheses, and formulate a decision rule.

You will be able to determine appropriate use of one- and two-tailed tests.

Key Terms

z test for a population mean -- A hypothesis test that evaluates how far the observed sample mean deviates, in standard error units, from the hypothesized population mean.

Null hypothesis (H_0) -- A statistical hypothesis that usually asserts that nothing special is happening with respect to some characteristic of the underlying population.

Alternative hypothesis (H_1) -- The opposite of the null hypothesis.

Research hypothesis -- Usually identified with the alternative hypothesis, this is the informal hypothesis or hunch that inspires the entire investigation.

Decision rule -- Specifies precisely when H_0 should be rejected (because the observed z qualifies as a rare outcome).

Critical z score -- A z score that separates common from rare outcomes and hence dictates whether H_0 should be retained or rejected.

Level of significance (a) -- Indicates the degree of rarity among random outcomes required to reject the null hypothesis (H_0).

Two-tailed or nondirectional test -- Rejection regions are located in both tails of the sampling distribution.

One-tailed or directional test -- Rejection region is located in just one tail of the sampling distribution.

Text Review

A null hypothesis is tentatively assumed to be true. It is tested by determining whether an observed sample mean qualifies as a common outcome or a rare outcome in the hypothesized sampling distribution. An observed sample mean qualifies as a (1)_____ (11.1) outcome if the difference betweeen its value and that of the hypothesized population mean is small enough to be viewed as merely another random outcome. A common outcome signifies that nothing special is happening in the underlying population and thus the null hypothesis should be (2)_____ (11.1). An observed sample mean qualifies as a (3)_____ (11.1) outcome if the difference between its value and the hypothesized value is too large to be reasonably viewed as merely another random outcome. A rare outcome will be a mean that deviates so far from the hypothesized mean that it would emerge from the sparse concentration of possible sample means in either tail of the sampling distribution.

For the actual hypothesis test, it is customary to convert the mean to (4)_____ (11.2), the familiar standard score conversion presented in Chapter 6. This conversion yields a sampling distribution that approximates the (5)_____ (11.2). The conversion is accomplished by the z score formula variation, formula 11.1, where z equals the observed sample mean minus the hypothesized population mean divided by the standard error.

The z test is accurate only when (1) the population is normally distributed or the sample size is large enough to satisfy the requirements of the (6)_____ theorem (11.2) and (2) the population standard deviation is known.

The most crucial and exciting phase of the research is the formulation of the (7)_____ (11.3).
The problem is then translated into the (8)_____ (11.3) hypothesis, which asserts that nothing special is happening with respect to some characteristic of the underlying population. The null hypothesis always makes a precise statement about a number, never a range of numbers. This single number actually used in the null hypothesis may be based on available information about a relevant population, or it may be based on some existing standard or theory. The null hypothesis also always makes a statement about a characteristic of the (9)_____ (11.4) never about a characteristic of the sample.

In general, the alternative hypothesis asserts the opposite of the (10)_____ (11.5) hypothesis and it specifies a range of values about the single number that appears in the null hypothesis. The alternative hypothesis is usually identified with the (11)_____ (11.5) hypothesis, the informal hypothesis or hunch that, by implying the presence of something special in the underlying population, serves as inspiration for the entire investigation.

A (12)_____ (11.6) specifies precisely when H_0 should be rejected. Decision rules are based on critical z scores that separate common from rare outcomes and dictate whether H_0 should be retained or rejected. We can identify the proportion of the total area under the sampling distribution that is identified with rare outcomes. This proportion is often referred to as (13)_____ (11.6). The level of significance indicates the degree of rarity among random outcomes required to reject the null hypothesis and is called (14)_____ (11.6). The null hypothesis is rejected if the observed z value exceeds the critical z value corresponding to a predetermined alpha because it deviates too far into the tails of the sampling distribution. After a decision has been made to retain or reject the null hypothesis, the decision must be interpreted.

Retaining H_0 is viewed as a weak decision and rejecting H_0 is seen as a (15)_____ (11.11) decision. The decision to retain H_0 implies not that it is probably true, but that it could be true, whereas the decision to reject H_0 implies that it is probably false and that H_1

is probably true. This is not a serious problem, as most researchers ultimately hope to reject the null hypothesis.

The research hypothesis is not tested directly because it lacks the necessary precision. A hypothesis must specify a single number about which the sampling distribution can be constructed. The null hypothesis meets this requirement. Furthermore, because the research hypothesis is identified with the alternative hypothesis, the decision to reject the null hypothesis will provide strong support for the research hypothesis.

In a two-tailed or (16)_____ (11.13) test, rejection regions are located in both tails of the sampling distribution, and the alternative hypothesis is concerned with a difference in population mean in either direction. The difference could be either higher or lower. In a one-tailed or (17)_____ (11.13) test, the rejection region is located in just one tail of the sampling distribution. Therefore, the observed sample mean triggers the decision to reject the null hypothesis only if it differs in the specified direction. Before the hypothesis test is conducted, the researcher must decide whether to conduct a one- or two-tailed test and which direction (if one-tailed) is significant. The (18)_____ test is extrasensitive.

The level of significance must also be chosen before the hypothesis test is conducted. The level of significance equals the probability that even though the null hypothesis is true, an error could occur and it could be rejected. Therefore, when the rejection of a true null hypothesis would have serious consequences, a smaller level of significance would be appropriate. Alpha may be set to equal .01 or .001. In real life, the researcher chooses alpha and must do so before looking at the data.

Problems and Exercises

1. Nationwide, the average age of nursing home residents is 76 years with a standard deviation of 3.5. A nursing home administrator in Texas wishes to determine whether the state average differs from the national average. Taking a random sample of 35

residents from nursing homes in Texas, he finds a
mean age of 78. Using alpha = .05, test the null
hypothesis. Be sure to write both hypotheses, a
decision rule, and an interpretation.

2. Average IQ is 100 with standard deviation of 15. An
    educational diagnostician wishes to determine
    whether learning-disabled students in her school
    district exceed the mean IQ. (Note: One of the
    criteria for learning-disabled students is having IQ
    in the normal or above normal range.) She takes a
    random sample of 33 students and finds a mean IQ of
    108. Using alpha = .05, test the null hypothesis.
    Be sure to write both hypotheses, a decision rule,
    and an interpretation.

3. The national average composite score on the ACT is 18 with a standard deviation of 5. The B.A. Nurse R.N. school uses a cutoff score of 16 as admission criteria. The program administrator wants to determine whether the current students' average is different from the national average of 18. A random sample of 42 currently enrolled students yields an average score of 20. Using alpha = .05, test the null hypothesis. Be sure to write both hypotheses, a decision rule, and an interpretation.

Beyond the Basics

In the library, find a journal article that reports the results of a z test. Determine whether the article format follows a structure similar to the one in your text for summarizing the hypotheses, the decision rule, critical z scores, and interpretation. What is different, if anything? Do you agree with the researcher's choice of one-tailed or two-tailed test? Do you agree with the choice of level of significance? Why or why not? What are your thoughts in regard to the sample size? Was it appropriate? Did sample size satisfy the central limit theorem requirements?

Post Test

1. Why are population means converted to z for hypothesis testing?

2. What are the assumptions that must be met in order to use a z test?

3. The _____ hypothesis supplies the value about which the hypothesized sampling distribution is centered.

4. The _____ asserts the opposite of the null hypothesis.

5. The rejection of H_0 is precisely stated by the _____.

6. The proportion of area under the curve that is identified with rare outcomes is referred to as _____ and is symbolized by _____.

7. The retention of the null hypothesis is viewed as a weak decision, but rejecting the null hypothesis is a strong decision. Explain why.

8. Exlain how a researcher knows whether to use a one-tailed or two-tailed test.

9. Unless there are obvious reasons for selecting a larger or smaller level of significance, a researcher would usually use alpha = _____.

10. Would a researcher be more likely to reject the null hypothesis with a one-tailed or two-tailed test?

Answers

Text Review

1. common
2. retained
3. rare
4. z scores
5. standard normal distribution
6. central limit
7. research problem
8. null
9. population
10. null
11. research
12. decision rule
13. level of significance
14. alpha
15. strong
16. nondirectional
17. directional
18. one-tailed

Problems and Exercises

1. $H_0: \mu = 76$
   $H_1: \mu \neq 76$

Reject H_0 at the .05 level of significance if z equals or is more positive than 1.96 or if z equals or is more negative than -1.96.

   $z = 3.39$

Reject H_0

Interpretation: The average age of Texas nursing home residents differs from the national average.

2. $H_0: \mu = 100$
   $H_1: u > 100$

Reject H_0 at the .05 level of significance if z equals or is more positive than 1.65.

   $z = 3.07$

Reject H_0

Interpretation: The average IQ of students in the local school district exceeds the national average.

3. H_0: $u = 18$
    H_1: $u \neq 18$

Reject H_0 at the .05 level of significance if z equals or is more positive than 1.96 or if z equals or is more negative than −1.96.

  $z = 2.60$

Reject H_0

Interpretation: The average ACT score of students at the B.A. Nurse R.N. school differs from the national average.

Post Test

1. Converting to z eliminates the original units of measure and standardizes the hypothesis testing procedure across all situations.
2. a. The population must be normally distributed or sample size must be large enough to meet the central limit theorem.
    b. The population standard deviation must be known.
3. null
4. alternative
5. decision rule
6. level of significance, alpha
7. The decision to retain H_0 implies not that H_0 is probably true, but that it could be true, the decision to reject implies that H_0 is probably false.
8. When the researcher is concerned that the true population mean differs from the hypothesized population mean only in a particular direction, the one-tailed test is used. For sensitivity to difference in either direction, use a two-tailed test.
9. .05
10. One-tailed, because the rejection region represents more area under the curve.

CHAPTER 12

MORE ABOUT HYPOTHESIS TESTING

Learning Objectives

You should develop knowledge and understanding of the key terms.

You will be able to differentiate between a type I and a type II error.

Key Terms

Type I error -- Rejecting a true null hypothesis.

Type II error -- Retaining a false null hypothesis.

Effect -- Any difference between a true and a hypothesized population mean.

Power (1-B) -- The probability of detecting a particular effect.

Power curves -- Cross-reference the likelihood of detecting any possible effect with different sample size.

Alpha (a) -- The probability of a type I error, that is, the probability of rejecting a true null hypothesis.

Beta (B) -- The probability of a type II error, that is, the probability of retaining a false null hypothesis.

Text Review

When conducting a hypothesis test, there are four possible outcomes. Rejecting a false null hypothesis would constitute a (1)_____ (12.2). Retaining a (2)_____ (12.2) null hypothesis would also be a correct decision. However, if the researcher rejects a true null hypothesis, he would make a (3)_____ (12.2). It would be a (4)_____ error (12.2) if the researcher retains a false null hypothesis. The null hypothesis is so important because it states that there is no (5)_____ (12.2), contradicting the research hypothesis.

When H_0 is true, the hypothesized sampling distribution qualifies as the (6)_____ (12.2) sampling distribution. However, when a randomly selected sample mean originates from the rejection region just by chance, then H_0 is rejected and the researcher has made a (7)_____ (12.3) error. The probability of this type I error equals (8)_____ (12.3). The probability of a correct decision equals (9)_____ (12.3).

Type I errors are often called (10)_____ (12.3) because decisions may be made, money may be spent, or further research may be prompted when none of these is truly warranted.

When H_0 is false, an incorrect decision or type II error is called (11)_____ (12.4) because the effect goes undetected. The probability of a type II error is (12)_____ (12.4). Whenever the effect is (13)_____ (12.4), the probability of a correct decision is high and equals (14)_____ (12.4). On the other hand, when the effect is (15)_____ (12.5), the probability of a correct decision is lower and the probability of a type II error increases.

One way to increase the probability of detecting a false null hypothesis is to increase (16)_____ (12.6). This is true because increasing sample size causes a reduction in (17)_____ (12.6). An extremely large sample size will thus produce a very sensitive hypothesis test. This is not always desirable because the test would detect even a small effect that has no practical importance. Appropriate sample size can be determined using (18)_____ (12.7).

The power of a hypothesis test equals the probability of detecting (19)_____ (12.8). To determine appropriate sample size, the researcher must decide (1) what is the smallest effect that merits detection? and (2) what is an appropriate detection rate? When these two questions have been answered, the researcher determines sample size by consulting (20)_____ (12.8).

Problems and Exercises

1. Before a hypothesis test, we are concerned about four possible outcomes. After a decision has been made to retain or reject the null hypothesis, we are concerned about only two possible outcomes. Explain.

2. A researcher reports that H_0 was rejected at the .05 level of significance for a hypothesis test using 250 subjects. Make suggestions as to how this research could be improved.

3. If H_0 is true, the probability of a type I error will always equal alpha. What is the probability of a correct decision?

4. If H_0 is false, the probability of a type II error is beta. What is the probability of a correct decision?

5. What can be done to increase the probability of detecting a false H_0?

6. What factor (that cannot be manipulated by the researcher) can increase the probability of a type II error or decrease the probability of a correct decision?

7. In the preceding question, why is effect described as a factor that the researcher cannot manipulate?

88

Beyond the Basics

Here is a crossword puzzle to help with key terms from Chapters 7-12.

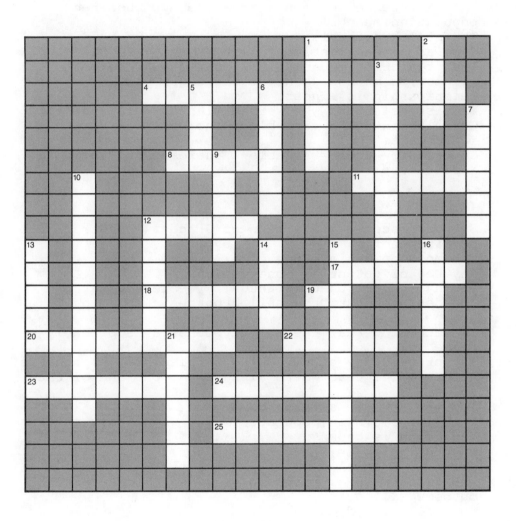

Across Clues

4. Type of z test where alpha is split to both ends of the curve
8. Level of significance
11. Subset of a population
12. Error of rejecting a true null hypothesis
17. Sample selected in such a way that every member of the population has an equal chance of being selected
18. The difference between a true and hypothesized population mean
19. The null hypothesis
20. Same as a nondirectional test
22. The prediction equation that serves as the best prediction method whenever the underlying relationship is linear.
23. British scientist who worked with correlation (r)
24. Correlation where variable X increases as variable Y increases
25. Correlation where variable X increases as variable Y decreases

Down Clues

1. Standard score with a mean of 0 and a standard deviation of 1
2. The most common average
3. Any set of potential observations
5. The hypothesis that is actually tested
6. The decision that the null hypothesis could be true
7. The decision that the null hypothesis is probably false
9. The probability of detecting a false null hypothesis
10. A one-tailed test where alpha is at one end of the curve
12. The error where a false H_O is retained
13. Test for a single population mean when the standard deviation is known
14. When H_O is false, the probability of a type II error
15. The proportion or fraction of times that a certain outcome will occur
16. The bell-shaped curve
21. A relationship between X and Y described by a straight line

Post Test

True or False

1. If H_0 is true, it is a correct decision to retain H_0.
2. A type I error consists of rejecting a true null hypothesis.
3. A type II error consists of retaining a true null hypothesis.
4. When generalizing beyond existing data, there is always the possibility of a type I or type II error.
5. If the null hypothesis is true, the probability of a type I error equals 1 - alpha.
6. If H_0 is false, the probability of a type II error is equal to beta.
7. The smaller the effect, the lower the probability of a type II error.
8. The probability of detecting a false null hypothesis can be increased by decreasing sample size.
9. A good way to determine appropriate sample size is to use a power curve.
10. A proper sample size is neither unduly small nor excessively large.

Answers

Text Review

1. correct decision
2. true
3. type I error
4. type II error
5. effect
6. true
7. type I
8. alpha
9. 1-alpha
10. false alarms
11. a miss
12. beta
13. large
14. 1-beta
15. small
16. sample size
17. standard error
18. power curves
19. an effect
20. power curves

Problems and Exercises

1. After the researcher has decided to retain H_0, he has either made a correct decision or a type II error. If the researcher decides to reject H_0, he has either made a correct decision or a type I error.

2. The researcher should reduce sample size or decrease alpha to .01, or both. As it is, the experiment may produce results that are statistically significant but of no practical importance.
3. 1-alpha
4. 1-beta
5. Increase sample size.
6. The size of the effect. The smaller the effect, the greater the probability of a type II error and the smaller the probability of a correct decision.
7. The researcher can alter sample size or determine alpha level, which affects probabilty of correct decisions or errors. But the size of the effect relies simply on the outcome of the experiment and cannot be manipulated.

Beyond the Basics

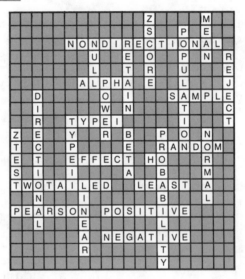

Post Test

1. true
2. true
3. false
4. true
5. false
6. true
7. false
8. false
9. true
10. true

CHAPTER 13

ESTIMATION

Learning Objectives

You should develop knowledge and understanding of the key terms.

You will be able to construct and interpret confidence intervals.

Key Terms

Point estimate -- A single value that represents some unknown population characteristic, such as the population mean.

Confidence interval -- A range of values that, with a known degree of certainty, includes an unknown population characteristic, such as a population mean.

Level of confidence -- The percent of time that a series of confidence intervals includes the unknown population characteristic, such as the population mean.

Text Review

A researcher may wish to estimate the value of a population mean rather than test a hypothesis based on a population mean. Estimation is possible with the use of (1)_____ and (2)_____ (13.1). A point estimate is a single value that represents some unknown (3)_____ (13.2) characteristic. The problem with point estimates is that they tend to be (4)_____ because of (5)_____ (13.2). Therefore, the researcher uses a more accurate type of estimate, (6)_____ (13.2). A confidence interval is a range of values that, with a known degree of certainty, includes an unknown population characteristic, such as a (7)_____ (13.3).

To understand confidence intervals, you must be aware of three important properties of the sampling distribution of the mean. (1) The mean of the sampling distribution always equals the (8)_____ (13.4). (2) The standard error of the sampling distribution equals the population standard deviation divided by the square root of (9)_____ (13.4). (3) The shape of the sampling distribution approximates a normal distribution if sample size satisfies (10) _____ (13.4).

A confidence interval can be constructed using formula 13.1, where a value of z from the standard normal table is multiplied by the standard error and this value is both added to and subtracted from the mean. In order to use this formula, the (11)_____(13.5) must be known and sample size must be at least (12)_____ (13.5).

In practice, only one confidence interval is actually constructed, and it is either (13)_____ or (14) _____ (13.6). Although we never really know whether a particular confidence interval is true or false, we can be reasonably confident when the level of confidence is 95 percent or more. The level of confidence indicates the percent of time that a series of confidence intervals includes the unknown population characteristic such as the population mean. An increase in confidence level causes a wider confidence interval that is less (15)_____ (13.7) unless it is offset by an increase in (16)_____ (13.7).

Although many different levels of confidence have been used, (17)_____ and (18)_____ (13.7) are the most prevalent. Confidence intervals are more narrow or precise when standard error is smaller. Since standard error is reduced when sample size is increased, the larger the sample size, the more precise the confidence interval.

In the behavior sciences, hypothesis tests have been preferred to confidence intervals. However, (19)_____ (13.9) are more informative because they indicate the size of (20)_____ (13.9). A hypothesis test is most appropriate when the major concern is whether or not (21)_____ (13.9) is present. Otherwise, a confidence interval could be used and should be considered when the null hypothesis is rejected.

Very large sample sizes result in small (22)_____ (13.10). The large sample size is not always desired in experiments, as explained in Chapter 12, but large samples are inexpensive and highly desirable in polls and surveys.

Problems and Exercises

1. In Chapter 11, problem 1, it was concluded that the
    average age of nursing home residents in Texas
    differs from the national average of 76. Given a
    standard deviation of 3 years and a sample mean of
    79, for a random sample of thirty-six elderly
    persons, construct a 95 percent confidence interval
    for mean age and interpret the confidence interval.

2. Refer to Chapter 11, problem 2, pg. 80. The null
    hypothesis was rejected. Now the educational
    diagnostician wishes to construct a confidence
    interval. Using the information in this problem
    from Chapter 11, answer the following questions.

    a. What is the best estimate of the unknown mean IQ
        for the entire population of learning-disabled
        students in the local school district?

    b. Construct a 99 percent confidence interval for
        the unknown population mean.

    c. Interpret the confidence interval.

3. On the basis of a random sample of 200 teenage viewers, a television pollster decides to cancel an after-school program series, reporting, with 95 percent confidence that only between 22 and 38 percent of the potential viewing audience have been watching the series.

    a. Comment on the width of the confidence interval.

    b. What could be done to make the confidence interval more precise?

Beyond the Basics

Think about the logic a researcher would apply in deciding whether to construct a 95 or 99 percent confidence interval. Essentially the decision is based on the magnitude of the consequences of making an error. Find two research reports in professional journals, one reporting a 95 percent confidence interval and one reporting a 99 percent confidence interval. Do you agree with the researcher's judgment in each case? What is there about the research that makes a false interval more critical?

Post Test

1. What are some of the similarities between hypothesis tests and confidence intervals?

2. Why are larger samples more acceptable in polls and surveys than in experiments?

3. What three properties of sampling distributions are essential to the understanding of confidence intervals?

4. What assumptions must be met in order to use formula 13.1 in constructing confidence intervals?

5. What is the relationship between sample size, standard error, and confidence intervals?

6. What is the difference between a point estimate and a confidence interval?

7. What is the disadvantage of using point estimates?

Answers

1. point estimates
2. confidence intervals
3. population
4. inaccurate
5. sampling variability
6. confidence intervals
7. population mean
8. population mean
9. the sample size
10. the central limit theorem
11. population standard deviation

12. 25
13. true
14. false
15. precise
16. sample size
17. 95
18. 99
19. confidence intervals
20. the effect
21. an effect
22. margins of error

Problems and Exercises

1. $\underline{X} \pm (\underline{z}\ conf)\ (\sigma_{\overline{X}}) =$

   $79 \pm (1.96)\ (.5) =$

   $79 \pm .98 = 78.02\ to\ 79.98$

   $\sigma_{\overline{X}} = \dfrac{\sigma}{\sqrt{n}} = \dfrac{3}{6} = .5$

   We can claim with 95 percent confidence that the interval between 78.02 and 79.98 includes the true population age.

2. a. 108

   b. $\underline{X} \pm (\underline{z}\ conf)\ (o_{x}) =$

   $108 \pm (2.58)\ (2.61) =$

   $108 \pm 6.73 = 101.27\ to\ 114.73$

   $\sigma_{\overline{X}} = \dfrac{\sigma}{\sqrt{n}} = \dfrac{15}{5.74} = 2.61$

   c. We can claim with 99 percent confidence, that the interval between 101.27 and 114.73 includes the true population IQ.

3. a. The confidence interval is too wide.
    b. To make the confidence interval more precise,
        switch to a lesser degree of confidence such as
        90 percent or increase the sample size.

Post Test

1. Both hypothesis tests and confidence intervals are
    more precise with larger samples. Both rely on
    values of z from the standard normal table. Both
    are based on properties of the sampling distribution
    listed for answer 3.
2. In experiments, the larger sample makes the
    hypothesis test more sensitive, so that the null
    hypothesis will almost always be rejected. This
    leads to the detection of an effect that may have
    statistical significance but no practical
    importance. Furthermore, experimentation may
    involve some very expensive and time-consuming
    procedures that would have to be applied to all
    members of the sample. In polls and surveys, the
    viewpoint is that the larger the sample is the
    better since increases are more likely to create
    samples that accurately reflect the characteristic
    of the population. Also, in polls and surveys, the
    information-gathering techniques provide easy and
    inexpensive answers to questions, thus foregoing the
    problems of time and expense faced by experimenters.
3. Three important properties of confidence intervals
    must be understood. (1) The mean of the sampling
    distribution always equals the population mean. (2)
    The standard error of the sampling distribution
    equals the population standard deviation divided by
    the square root of the sample size. (3) The shape
    of the sampling distribution approximates a normal
    distribution if sample size satisfies the central
    limit theorem.
4. The use of formula 13.1 assumes that the population
    standard deviation is known and that the population
    is normal or that sample size is sufficiently large
    to meet the requirements of the central limit
    theorem.
5. As sample size increases, standard error decreases,
    and the confidence interval becomes more precise.
6. A point estimate specifies a single value for the
    unknown population mean, whereas the confidence
    interval supplies a range of values.
7. Point estimates are inaccurate because they do not
    take into account sampling variability.

CHAPTER 14

t TEST FOR ONE SAMPLE

Learning Objectives

You should develop knowledge and understanding of the key terms.

You will be able to identify research problems appropriate for the use of the t test for one sample.

You will be able to write hypotheses, solve for t, make correct decisions about the null hypothesis, and interpret the results of t tests.

Key Terms

Sample standard deviation (s) -- The version of the sample standard deviation, with n - 1 in its denominator, that is used to estimate the unknown population standard deviation.

Estimated standard error of the mean ($s_{\bar{x}}$) -- The version of the standard error of the mean that is used whenever the unknown population standard deviation must be estimated.

t ratio -- A replacement for the z ratio whenever the unknown population standard deviation must be estimated.

Degrees of freedom (df) -- The number of values free to vary, given one or more mathematical restrictions.

Text Review

The sampling distribution of t was discovered by (1) _____ (14.2). In reality, there is a family of t distributions, each associated with a number referred to as (2) _____ (14.2). In the present case, the number of degrees of freedom (df) always equals (3) _____ (14.2).

All t distributions are similar to z distributions in that they are symmetrical, unimodal, and bell shaped. The biggest difference between t and z distributions is the (4) _____ (14.2) of the t distribution.

Tables for t distributions contain only the values that correspond to the common levels of significance. The t values listed in the table are (5)_____ (14.3) and originate from the (6)_____ (14.3) half of the distribution. The symmetry allows the critical t values for the lower half of the distribution to be obtained by simply placing a negative sign in front of any table entry.

Since the population standard deviation is unknown, it must be estimated. The old formula for calculating standard deviation was appropriate for (7)_____ but not for (8)_____ statistics (14.4). The population standard deviation can be estimated from the sample standard deviation by replacing n with (9)_____ (14.4) in the denominator. When the unknown population standard deviation must be estimated, the population standard error of the mean must also be estimated. The shift to this estimate requires that the z test be changed to t, where each distribution has its own degrees of freedom.

The t distribution has greater variability than the z distribution. This increased variability arises from the estimated standard error of the mean and the chance differences that could occur in the estimates. The extra variability also explains the inflated tails of the curve in the t distribution. Because of this extra variability, one could expect actual critical t values to be (10)_____ (14.5) than critical z values.

To construct confidence intervals for estimating the population mean based on a t distribution, use the formula for confidence intervals already presented and substitute t for z. The formula reads $X + (t \text{ conf}) (s_{\bar{x}})$. The symmetrical limits of confidence intervals requires specifications like those of a two-tailed hypothesis test.

Use of the t test requires some basic assumptions. Use t rather than z when the (11)_____ (14.7) is unknown. Assume that the underlying population is (12)_____ (14.7). If the assumption of normality is violated, the accuracy of the test is relatively unaffected as long as (13)_____ (14.7) is sufficient.

(14)_____ (14.8) refers to the number of values that are free to vary, given one or more mathematical restrictions on the entire set of values. The mathematical restriction relevant to degrees freedom in t distributions is that the sum of all values, expressed as deviations from their mean, always equals (15)_____ (14.8).

All hypothesis tests represent variations on a common theme: If some observed characteristic, such as the mean for a random sample, qualifies as a rare outcome under the null hypothesis, the hypothesis will be (16)_____ (14.9). Otherwise, the hypothesis will be (17)_____ (14.9). To determine whether an outcome is rare, the observed characteristic is converted to a new value, such as t, and compared to critical values from the appropriate sampling distribution.

Problems and Exercises

1. A psychology professor believes that his summer school class grades exceed the class average of 70 usually earned by students in the fall and spring semester. A random sample of summer student grades yields the following results: 88, 72, 84, 77, 96, 67, 65.

   a. Test the null hypothesis with t using the .05 level of significance.

   b. Construct a 95 percent confidence interval for the true grade average.

   c. Interpret the confidence interval.

2. Find the critical t values for the following hypothesis tests:

   a. Two-tailed test, a = .001, df = 30

   b. One-tailed test, upper tail critical, a = .01, df = 21

   c. One-tailed test, lower tail critical, a = .05, df = 8

3. Now assume a researcher made an error in choosing the appropriate hypothesis test and used z instead of t in the hypothesis tests in problem 2.

   a. Find the critical z value for each test.

   b. How do the values differ?

   c. What causes the differences?

4. Behavioral psychologists have established that a fixed ratio reinforcement schedule produces the highest rate of response. Upon learning this, a manufacturer of blue jeans decides to apply the principle in one of the factories on a trial basis. Before the trial, the average seamstress sewed 21 pockets per hour and was being paid by the hour. During the trial period, seamstresses were paid by the number of pockets sewed (a fixed ratio reinforcement). The following results were obtained for a random sample of seamstresses: 18, 25, 28, 20, 31, 29, 24.

   a. Test the null hypothesis with t, using the .05 level of significance.

   b. Construct a 95 percent confidence interval for the true average number of pockets sewn in one hour.

   c. Interpret the confidence interval.

   d. If you owned stock in this company, would you favor the use of a fixed ratio reinforcement schedule to pay employees?

Beyond the Basics

In this section in Chapter 11, you were asked to find a journal article reporting a z test. If you did this, using the data in that article, pretend that a t test is appropriate and compare the appropriate critical t value to the critical z. Would the decision to reject or retain the null hypothesis be the same? Does the role of sample size change at all? If so, how?

If you did not find an article for the assignment in Chapter 11, you may now find one reporting either z or t. Then answer the preceding questions comparing the results against a critical value for both t and z.

Post Test

1. What determines whether a t test or z test should be used?

2. How do critical values of t compare with critical
    values of z?

3. What are the assumptions that must be met for the
    proper use of a t test?

4. What happens to the t test if the assumption of
    normality is violated?

5. Define degrees of freedom.

Answers

Text Review

1. William Gossett
2. degrees of freedom
3. sample size minus one
4. inflated tails
5. positive
6. upper
7. descriptive
8. inferential
9. n - 1

10. larger
11. population
     standard deviation
12. normally distributed
13. sample size
14. degrees of freedom
15. zero
16. rejected
17. retained

Problems and Exercises

1. a. using a one-tailed test, upper tail critical,
      t = 1.95;   reject the null hypothesis

   b. The 95 percent confidence interval = 67.86 –
      89.0

   c. It can be claimed with 95 percent confidence
      that the interval between 67.86 and 89.0
      includes the true mean grade for summer
      students.

2. a. 3.646   b. 2.518   c. 1.860

3. a. 3.30   b. 2.33   c. 1.65
      z values are smaller than t values because
      of inflated tails of the t distribution.

4. a. using a one-tailed test, upper tail critical
      t = 2.22;   reject the null hypothesis
   b. The 95 percent confidence interval equals
      21.5 – 28.5.
   c. It can be claimed with 95 percent confidence
      that the interval between 21.5 and 28.5 includes
      the true mean number of pockets sewed.
   d. Yes, because production is increased when the
      fixed ratio reinforcement schedule is used.

Post Test

1. A t test should be used when the population
   standard deviation is unknown.
2. The inflated tails of the t distribution cause t
   values to be larger than z values, especially if
   sample size is small.
3. To use t, you must assume that the underlying
   population is normally distributed.
4. If the assumption of normality is violated, the t
   test will retain most of its accuracy as long as
   sample size is sufficiently large (larger than ten).
5. Degrees of freedom refers to the number of values,
   within a set of values, that are free to vary, given
   some mathematical restriction. Specifically, in a
   t test, degrees of freedom is related to the
   estimate of the population standard deviation and
   the use of n – 1 in the denominator of the formula
   for that estimate.

$\underline{t}$ FOR TWO INDEPENDENT SAMPLES

Learning Objectives

You should develop knowledge and understanding of the key terms.

You will be able to identify research problems appropriate for the use of the $\underline{t}$ test for two independent samples.

You will be able to write hypotheses, solve for $\underline{t}$, make correct decisions about the null hypothesis, and interpret the results of $\underline{t}$ tests for two independent samples.

Key Terms

Two independent samples -- Observations in one sample are not paired, on a one-to-one basis, with observations in the other sample.

Sampling distribution of $\overline{X}_1 - \overline{X}_2$ -- Differences between sample means based on all possible pairs of random samples.

Standard error of $\overline{X}_1 - \overline{X}_2$ ($s_{\overline{X}_1 - \overline{X}}$) -- A rough measure of the average amount by which any difference between sample means deviates from the difference between population means.

Pooled variance estimate (s^2_p) -- The most accurate estimate of the variance (assumed to be the same for both populations) based on a combination of two sample variances.

Confidence interval for $\mu_1 - \mu_2$ -- A range of values that, in the long run, includes the unknown difference between population means a certain percent of the time.

Statistical significance -- An indication that is not of practical importance; it indicates merely that the null hypothesis is probably false.

p-value -- The degree of rarity of a test result, given that the null hypothesis is true.

107

Text Review

Two independent samples occur when (15.2) (1)_____ in one sample are not paired with observations in the other sample. When a t test is conducted for two independent samples, the difference between population means reflects the (2)_____ (15.3) of the variable being studied. In the example in the text, the variable is (3)_____ (15.3). When there is little difference between the two population means, there is little effect.

The null hypothesis states there is (4)_____ (15.4) difference between population means. There are three possible alternative hypotheses. One states that the difference between population means doesn't equal zero. This would represent a (5)_____ (15.4) test. A second possible hypothesis states that the difference is less than zero. This is a one-tailed test with the (6)_____ (15.4) tail critical. The third possibility is a one-tailed test with (7)_____ (15.4) tail critical which states that the difference between population means (8)_____ (15.4) zero. When there is concern only about differences in a particular direction, a (9)_____ or one-tailed hypothesis test should be used.

Just as the sampling distribution of the mean (presented in Chapter 10) is not actually constructed, the sampling distribution for the difference between sample means is not constructed. Instead, we rely on statistical theory to provide information about the mean and standard error for the sampling distribution of $\bar{X}_1 - \bar{X}_2$. In practice, there is only one observed difference and the (10)_____ (15.5) is conducted to determine whether it qualifies as a common or rare outcome. In the one-sample case, the mean of the sampling distribution equals the mean of the (15.6) (11)_____. In the two-sample case, the mean of the sampling distribution equals the (12)_____ (15.6) between population means.

The sampling distribution of $\bar{X}_1 - \bar{X}_2$ has a standard deviation referred to as the (13)_____ (15.7) of the difference between sample means. The standard error is a rough measure of the average amount by which any difference between sample means deviates from the difference between (14)_____ (15.7). The size of the standard error decreases as sample size (15)_____ (15.7).

Before the t ratio can be calculated, the standard error must be estimated. The t test assumes that the two population variances are (16)_____ (15. 10). The pooled variance estimate can be obtained by combining the variance common to both populations. The pooled variance estimate is the most accurate estimate of the variance (assumed to be the same for both populations) based on a combination of the two sample variances. The degrees of freedom for the pooled variance estimate equal the sum of the two sample sizes minus two. Two degrees of freedom are lost because the (17)_____ (15.10) in each of the two samples are expressed as (18)_____ (15.10) from their respective sample means.

A confidence interval may be constructed for the difference between population means. The confidence interval is a range of values that, in the long run, includes the unknown difference between population means a certain percent of the time. If both positive and negative values appear in a confidence interval, no single interpretation is possible. The inclusion of a zero value in the range indicates that the variable being studied may have (19)_____ (15.12).

The t test for two independent samples assumes that both underlying populations are (20)_____ (15.13) and have (21)_____ (15.13) variances. If sample sizes are (22)_____ and _____ (15.13), violations of these assumptions will be of little concern.

Results of hypotheses tests are often described as having statistical significance if the null hypothesis has been rejected. Statistical significance indicates that the null hypothesis is probably false, but doesn't indicate whether it is seriously false or mildly false. Statistical significance may lack practical importance because sample size is (23)_____ (15.14). Practical importance can be gauged by checking the size of the observed effect relative to the size of the standard deviation. If the size of the effect is small (about one tenth the size of the standard deviation), the statistically significant effect probably lacks practical importance.

If the researcher does not retain or reject the null hypothesis, but views it with suspicion, depending on the degree of rarity of the test result, then (24)_____

(15.15) are being used. Smaller $\underline{p}$ values tend to discredit the null hypothesis and support the research hypothesis. Using $\underline{p}$ values is a less structured approach to hypothesis testing. One disadvantage of this approach is that when a firm decision is not being made to retain or reject the null hypothesis, it is difficult to deal with the concepts of type I and type II errors.

Level of significance is indicated before the test result is obtained and $\underline{p}$ values are indicated after. However, any $\underline{p}$ value less than .05 implies that the null hypothesis would have been (25)_____ (15.15) and any $\underline{p}$ value greater than .05 implies that the null hypothesis would be (26)_____ (15.15).

Problems and Exercises

1. A researcher wishes to determine the effect listening to music has on memory of word lists. Twelve student volunteers are randomly assigned to two groups—one group listens to music while studying the word list ($\underline{X}_1$) and one group does not ($\underline{X}_2$). The mean performance on the test of memory for $\underline{X}_1$ is 18 and the mean for $\underline{X}_2$ is 23. The estimated standard error equals 1.45. Use $\underline{t}$ to test the null hypothesis at the .05 level of significance.

2. A social psychologist wishes to determine the effect of fear on affiliation. Eighteen volunteers are randomly assigned to either a frightening description of an experiment or a nonfrightening description. Then the volunteers respond to a questionnaire as to their preference for waiting while the experiment is prepared. Higher scores indicate more desire for affiliation or waiting with another volunteer. Lower scores indicate a preference for waiting alone. Using t, test the null hypothesis at the .01 level of significance.

Affiliation Scores	
Frightened	Not Frightened
18	12
15	6
17	8
13	13
15	12
19	6
22	10
19	7
20	11

3. Problem 6 from your textbook reads as follows. An investigator wishes to determine whether alcohol consumption causes a deterioration in the performance of automobile drivers. Before the driving test, subjects drank a glass of orange juice, which in the case of one group of subjects was laced with two ounces of vodka. Performance was measured by the number of errors made on a driving simulator, such as those found in many amusement park arcades. One hundred twenty volunteer subjects were randomly assigned, in equal numbers, to the two groups. For the subjects who drank the laced orange juice, the mean number of errors ($\bar{X}_1$) equaled 26.4, and for the subjects who drank the regular orange juice, the mean number of errors ($\bar{X}_2$) equaled 18.6. The estimated standard error equaled 2.4. To understand the effect of sample size, assume that the original group of volunteers consisted of only thirty drivers divided equally into two groups. This would increase the estimated standard error. Let's assign the estimated standard error a new value of 3.9. Now use t to test the null hypothesis at the .05 level of significance. Is the decision the same? What is the effect of sample size on the outcome of the test?

Beyond the Basics

Make-up your own research problem for a _t_ test for two independent samples. Write it up as a problem so that your fellow students can find the solution. Then work out a solution following the format in Table 15.1 in your text.

Post Test

1. List two important properties of the sampling distribution of the difference between sample means.

2. What is the appropriate _degrees of freedom_ for the _t_ distribution for two independent samples?

3. When does the confidence interval for the difference between population means have a single interpretation?

4. What are the assumptions for the use of a _t_ test?

5. What is the difference between statistical significance and practical importance?

6. What is the difference between level of significance
    and p value?

Answers

<u>Text Review</u>

1. observations
2. effect
3. blood doping
4. no
5. nondirectional
6. lower
7. upper
8. exceeds
9. directional
10. hypothesis test
11. population
12. difference
13. standard error
14. population means
15. increases
16. equal
17. observaitons
18. deviations
19. no effect
20. normally distributed
21. equal
22. fairly large and equal
23. excessively large
24. p values
25. rejected
26. retained

<u>Problems and Exercises</u>

1. Using a nondirectional or two-tailed test
    t = 3.45; reject the null hypothesis
2. Using a nondirectional or two-tailed test
    t = 6.17; reject the null hypothesis
3. Using a one-tailed test, upper tail critical
    t = 2.00; retain the null hypothesis; In this
    case, the decrease in sample size caused the null
    hypothesis to be retained instead of rejected
    because of the increase in the estimated standard
    error.

114

1. a. The mean of the sampling distribution equals the difference between population means.
    b. The standard error roughly measures the average amount by which any difference between sample means deviates from the difference between population means.
2. Degrees of freedom is $n_1 + n_2 - 2$.
3. When the two limits of the interval share similar signs. Both are either positive or negative.
4. Both underlying populations are normally distributed with equal variances.
5. Statistical significance is indicated by the rejection of the null hypothesis. When sample size is too large, this may occur even though the research has no practical importance or "real world" application.
6. Level of significance must be determined before the data is analyzed and is the basis for rejecting or retaining the null hypothesis. The p value is determined after the hypothesis test and indicates the degree of rarity of the test result, given that the null hypothesis is true.

t FOR TWO DEPENDENT SAMPLES

Learning Objectives

You should develop knowledge and understanding of the key terms.

You will be able to identify research problems appropriate for the use of the t test for two dependent samples.

You will be able to write hypotheses, solve for t, make correct decisions about the null hypothesis, and interpret the results of t tests for two dependent samples.

Key Terms

Dependent samples -- Each observation in one sample is paired, on a one-to-one basis, with a single observation in the other sample.

Difference score -- The arithmetic difference between each pair of scores in two dependent samples.

Repeated measures -- Whenever the same subject is measured more than once.

Counterbalancing -- Reversing the order of conditions for half of all subjects.

Text Review

Two dependent samples occur when each observation in one sample is (1)_____ (16.2) with a single observation in the other sample. The t test for dependent samples involves using the difference score. This converts the two sample problem to a (2)_____ (16.3) problem. Thus, the original pair of populations becomes a single population of (3)_____ (16.4). If the variable being tested has no effect, then the population mean of all difference scores should equal (4)_____ (16.4). A directional hypothesis with upper tail critical would state that the mean difference (5)_____ (16.4) zero.

A directional hypothesis with lower tail critical would state that the mean difference is (6)_____ (16.4) than zero. A nondirectional hypothesis simply states that the mean difference is different from or not equal to zero. The mean of the sampling distribution of $\overline{D}$ equals the difference between population means.

Degrees of freedom is different for the $\underline{t}$ test for two dependent samples because $\underline{n}$ equals the number of (7) _____(16.5) and not the number of observations.

One advantage of matching is that it eliminates one source of (8)_____ (16.6). Confidence intervals are interpreted much the same for dependent samples as for independent samples.

Matching is desirable only when some uncontrolled variable appears to have a considerable impact on the variable being measured. Appropriate matching reduces (9)_____ (16.9) and the estimated (10)_____ (16.9). Therefore, matching (11)_____ (16.9) the sensitivity of the hypothesis test. To identify a variable that should be matched, the researcher should be familiar with all previous research in the area and conduct (12)_____ (16.9).

A test could involve the use of the same subjects in both samples. This is referred to as (13)_____ (16.10). A big advantage of repeated measures is that uncontrolled variability due to individual differences is eliminated. This advantage may be offset by other concerns, such as fatigue, practice, or motivation when subjects must perform twice.

When subjects must perform in both conditions, it is customary to use (14)_____ (16.9), reversing the order of the conditions for half the subjects.

The present $\underline{t}$ test assumes that the population of difference scores is normally distributed. Violations of this assumption will not matter much as long as (15)_____ (16.11) is sufficiently large.

When the researcher wishes to determine whether there is a relationship between two variables, the appropriate measure is the $\underline{t}$ test for the population correlation coefficient. The degrees of freedom is $\underline{n}$ - 2. Two

for variable $\underline{Y}$. In the $\underline{t}$ test for population correlation coefficient, sample size must be sufficiently large to minimize sampling variability. In using this test, the researcher must assume that the sample originates from a normal (16)_____ (16.14) population.

Problems and Exercises

1. A random sample of 45 applicants for a Master's program in clinical psychology reveals an $\underline{r}$ of .53 between their Graduate Record Exam scores and their grade point average in the coursework. Test the null hypothesis with $\underline{t}$, using the .05 level of significance.

118

2. An investigator wants to examine the effects of alcohol consumption on the number of errors made on a measure of manual dexterity. Four of the volunteers are tested first without alcohol, X_1, and then retested after consuming two drinks, X_2. The other five are tested in the reverse order with time allowed for the alcohol effects to wear off. Their performance is indicated in the following table.

Number of Errors on Manual Dexterity

Subject	X_1	X_2
1	2	6
2	1	4
3	4	5
4	3	7
5	1	5
6	2	5
7	3	6
8	3	8
9	2	4

a. Using t, test the null hypothesis at the .05 level of significance.

b. Specify the p-value for the test result.

3. Assume the data in the following table were observed from an actual study represented by problem 9a. in your text. A researcher wishes to determine whether attendance at a day-care center increases the scores of three-year-old children on a motor skills test. Random assignment dictates which member from each of 10 pairs of twins attends the day-care center and which member stays at home.

Effect of Day-care on Motor Skills		
Pair No.	Day-care	At Home
1	23	20
2	19	16
3	17	18
4	25	20
5	22	19
6	24	18
7	21	20
8	18	18
9	20	17
10	25	19

a. Using t, test the null hypothesis at the .05 level of significance.

b. Specify the p-value for the test result.

c. What makes the samples dependent?

d. What are the variables important to control? How does the design of the experiment control them?

Beyond the Basics

Make up your own research problem for a _t_ test for two dependent samples. Write it up as a problem so that your fellow students could find the solution. Then work out a solution following the format in Table 16.1. Compare the make-up of the samples used in this problem with the samples you created for this section in Chapter 15. How are dependent samples different from independent samples? Would it be appropriate to use the same research question you used before and alter the samples? Why or why not?

Post Test

1. What significant transformation takes place with the use of difference scores?

2. What is a major difference between the sampling distribution for the difference scores and the sampling distribution for independent samples?

3. What is the appropriate degrees of freedom for the _t_ distribution of two dependent samples?

4. What are the disadvantages of matching?

5. What are the advantages of matching?

6. When is matching appropriate?

7. What is repeated measures?

8. What are the assumptions for the t test for dependent samples?

9. What happens if the assumption is violated?

10. What are the assumptions for using the <u>t</u> test for correlation coefficients?

Answers

Text Review

1. paired
2. one sample
3. difference scores
4. zero
5. exceeds
6. less
7. difference scores
8. variability
9. variability
10. standard error
11. increases
12. pilot studies
13. repeated measures
14. counterbalancing
15. sample size
16. bivariate

Problems and Exercises

1. $t = 4.11$; reject the null hypothesis
2. Using a two-tailed test,
   $t = -8.043$; reject the null hypothesis $p < .001$
3. Using a one-tailed test, upper tail critical
   $t = 3.856$; reject the null hypothesis $p < .01$
   The samples are dependent because we could think of identical twins as being matched on all hereditary characteristics, since they have the same genetic makeup. Motor ability, age, and experience might all be important variables that would be controlled by using the identical twins who would likely have the same experiences from similar environments as well as the identical genes.

Post Test

1. The original pair of populations is converted to a single population.
2. The standard error is smaller when the two samples are dependent.
3. $n - 1$
4. Subjects may be lost. It is costly and time-consuming.

123

5. It reduces the size of the estimated standard error.
6. Matching is appropriate only when an uncontrolled variable has been identified that aids in the interpretation of the preliminary findings.
7. Repeated measures is a special case where the same subjects are used in both samples.
8. The assumption is that the population of difference scores is normally distributed.
9. The result is relatively unaffected if sample size is sufficiently large.
10. The assumptions are that the population distributions for X and Y are normally distributed and the relationship between X and Y is linear.

ANALYSIS OF VARIANCE (ONE WAY)

Learning Objectives

You should develop knowledge and understanding of the key terms.

You will identify research problems appropriate for use of ANOVA.

You will be able to write hypotheses, solve for F, make correct decisions about the null hypothesis, and interpret the results of F tests for one-way ANOVA.

You will identify the appropriate situations for use of Scheffe's test and be able to interpret the results.

Key Terms

Analysis of variance (ANOVA) -- An overall test of the null hypothesis for three or more population means.

Treatment effect -- At least one difference between population means defined for the various experimental conditions.

Variability between groups -- Variability among scores of subjects who, being in different groups, receive different experimental treatment.

Mean Square (MS) -- A variance estimate in ANOVA consisting of a sum of squares divided by its degrees of freedom.

Degrees of freedom (df) -- The number of deviations free to vary in any sum of square term.

Multiple comparisons -- The series of possible comparisons whenever, as in ANOVA, more than two population means are involved.

One-way ANOVA -- The simplest type of anlaysis of variance where population means differ only with respect to one dimension or factor.

Random error -- The combined effects (on the score of individual subjects) of all uncontrolled factors.

Variability within groups -- Variability among scores of subjects who, being in the same group, receive the same experimental treatment.

Sum of squares (SS) -- The sum of the squared deviations of some set of scores about their mean.

F ratio -- Ratio of the between-group mean square (for subjects treated differently) to the within-group mean square (for subjects treated similarly.)

Scheffe's test -- A multiple comparison test that, regardless of the number of comparisons, never permits the cumulative probability of at least one type I error to exceed the specified level of significance.

Text Review

Testing the null hypothesis for three or more population means requires a statistical procedure known as (1) _____ (17.1). Specifically, Chapter 17 deals with (2)_____ (17.1) ANOVA, where population means differ with respect to only one factor. One source of variability in ANOVA is the differences between groups means. Small differences can be attributed to (3)_____ (17.2). However, relatively large differences between group means probably indicate that the null hypothesis is (4)_____ (17.2). If there is at least one difference between the population means, there is (5)_____ (17.2). A second source of variability in ANOVA is an estimate of the variability within groups (subjects treated similarly). To make a decision about the null hypothesis, these two sources of variability are compared. The more that the variability between groups exceeds the variability within groups, the more likely the null hypothesis will be false. Regardless of whether the null hypothesis is true or false, the variability within groups reflects only (6)_____ (17.2). Random error is the combined effects of all uncontrolled factors such as individual differences among subjects, variations in experimental conditions, and measurement errors. The within-group variability estimate is often referred to as the (7)_____ (17.2).

For three or more samples, the null hypothesis is tested with the $\underline{F}$ ratio, variability between groups divided by variability within groups. $\underline{F}$ has its own family of sampling distributions, so an $\underline{F}$ table must be consulted to find the critical $\underline{F}$ value. In the $\underline{F}$ test, if the variability between groups exceeds the variability within groups, then the null hypothesis will be rejected. If the null hypothesis is true, then the two estimates of variability (between and within groups) will reflect only random error. The values will be similar, so the $\underline{F}$ will be small and the null hypothesis will be retained.

Variance is a measure of (8)_____ (17.5). A variance estimate indicates that information from a sample is used to determine the unknown variance for a population. In ANOVA, a variance estimate is composed of the numerator, the sum of squares, and the denominator, which is always (9)_____(17.5). When the ratio is calculated, it produces the mean of the squared deviations referred to as (10)_____(17.5).

In ANOVA, most of the computational effort is in finding the various sum of squares terms. $SS_{between}$ equals the sum of the squared deviations of group means about their mean, the overall mean. SS_{within} equals the sum of the squared deviations of all scores about their respective group means. SS_{total} equals the sum of the squared deviations of all scores about the overall mean. Calculations of the sum of squares terms can be verified by calculating all three from scratch and then checking because the sum of squares total equals the sum of squares within and the sum of squares between added together. For each sum of squares term, degrees of freedom differ. In ANOVA, the degrees of freedom for sum of squares total always equals the combined degrees of freedom for the other sum of squares terms.

Mean squares between reflects the variability between groups who are treated (11)_____ (17.8). Mean squares within reflects the variability among scores for subjects who are treated (12)_____ (17.8). Mean squares within measures only (13)_____, but mean squares between measures (14)_____ (17.8).

The observed $\underline{F}$, once calculated, may be compared with the critical $\underline{F}$ specified by the pair of degrees of freedom associated with it. The critical $\underline{F}$ values are found in Table C, Appendix D.

Rejection of the null hypothesis indicates only that not all population means are equal. In order to pinpoint the one or more differences between pairs of population means that contribute to the rejection, a test of (15)_____ (17.11) must be used. Multiple t tests cannot be used because it would increase the probability of a (16)_____ (17.11) error. Once the overall null hypothesis has been rejected in ANOVA, Scheffe's test can be used for all possible comparisons without the probability of the type I error exceeding the (17) _____ (17.12). When sample sizes are unequal, Scheffe's critical value must be calculated for each comparison. But when sample sizes are equal, the critical mean difference is only calculated once and then used to evaluate the remaining comparisons. It is important to note that Scheffe's test should be used only when the overall null hypothesis has been rejected.

The F test in ANOVA is equivalent to a nondirectional test even though the rejection region appears only in the upper tail of the distribution. This is due to the squaring of all the values which makes it impossible to have a negative value for F.

The assumptions for the F test are the same as for t. All underlying populations are assumed to be normally distributed with equal variances. Violations of the assumptions are not critical as long as sample size is greater than 10. ANOVA techniques used in the text presume that scores are independent. Furthermore, attention should be paid to sample size so that it is not unduly small or excessively large.

Problems and Exercises

1. The principal of an elementary school wishes to determine the most effective method to improve reading as measured by the district's reading achievement test. Twenty-eight children in third grade are randomly assigned to one of three reading improvement programs. Test the difference between their September reading scores and April reading scores using ANOVA to determine whether any one method is superior. Use the .05 level of significance. The difference scores follow.

    * Don't forget the Scheffe's test if appropriate.

Tutoring	Home Help	Reading Aloud Practice
10	8	6
7	6	9
12	4	5
9	4	7
13	9	7
11	7	6
10	8	8
14	5	5
11	7	9
	8	

2. A medical researcher wishes to identify the best method for reducing blood pressure. Thirty-eight volunteers with high blood pressure are randomly assigned to each of four groups. After four months of observation and treatment, the following results are obtained. Results represent the difference in blood pressure before and after treatment. Using the F test, .05 level of significance, test the null hypothesis. Use Scheffe's test if applicable.

No treatment (Control)	Medication	Exercise	Meditation
0	15	10	12
2	10	11	9
1	12	9	7
3	14	7	9
2	13	11	11
5	16	12	8
3	11	10	10
4	13	7	11
1	14	10	6
	17		8

3. In the preceding problem, what is the purpose of having a group that receives no treatment?

Beyond the Basics

Contemplate a research question that could best be answered by using ANOVA. What are some of the criteria you would want to meet? How many groups would be in your study? What treatment measures would you use? How would you select the participants? Would there be a control group? What variables might contribute to random error? What would sample size be? What should the level of significance be and when should it be established? Are there other questions you should ask yourself in preparing your study? Write a research proposal including the answers to these and any other pertinent questions that you develop.

Post Test

1. In ANOVA, when the calculated F ratio yields a value close to 1, what can be concluded?

2. If the null hypothesis is false, what is the nature of the relationship between variability between groups and variability within groups?

3. What is indicated by the relationship described in question 2?

4. How are differences between specific pairs of population means pinpointed?

5. When is it appropriate to use Scheffe's test?

6. Why is the F test always nondirectional?

7. What are the assumptions that should be met in order to use the F test?

8. What would be the result of violating the assumptions of the F test?

9. In using Scheffe's test, what is the importance of sample size?

Answers

Text Review

1. analysis of variance
2. one-way
3. chance
4. false
5. treatment effect
6. random error
7. error term
8. variability
9. degrees of freedom

10. mean square
11. differently
12. similarly
13. random error
14. treatment effect
15. multiple comparisons
16. type I
17. level of significance

Problems and Exercises

1. Which of the three methods tested is the most effective in improving reading?

   Statistical hypotheses:
   H_0: $\mu_{tutoring}$ = $\mu_{home\ help}$ = $\mu_{reading\ aloud}$

   H_1: H_0 is false.

   Decision rule:
   Reject H_0 at the .05 level of significance if F equals or is more positive than 3.38, given $df_{between}$ = 2 and df_{within} = 25.

   F = 15.107

   Decision: Reject H_0

   Interpretation:
   There is evidence that one or more of the reading methods improves reading more than the others.

Source	SS	df	MS	F
Between	100.12	2	50.06	15.11
Within	82.844	25	3.31	
Total	182.964	27		

Scheffe's test may be used.
$\bar{X}_1 - \bar{X}_2$ = significant difference
$\bar{X}_1 - \bar{X}_3$ = significant difference
$\bar{X}_2 - \bar{X}_3$ = not a significant difference

2. Which of the four methods tested is the most effective in reducing blood pressure?

Statistical hypotheses:
H_0: $\mu_{control} = \mu_{medication} = \mu_{exercise} = \mu_{meditation}$

H_1: H_0 is false.

Decision rule:
Reject H_0 at the .05 level of significance if F equals or is more positive than 2.88, given $df_{between}$ = 3 and df_{within} = 34.

$F = 57.42$

Decision: Reject H_0

Interpretation:
There is evidence that one or more of the methods is better at reducing blood pressure.

Source	SS	df	MS	F
Between	604.92	3	201.64	57.42
Within	119.40	34	3.51	
Total	724.32	37		

Scheffe's test may be used.

$\bar{X}_1$ vs $\bar{X}_2$ = significant difference
$\bar{X}_1$ vs $\bar{X}_3$ = significant difference
$\bar{X}_1$ vs $\bar{X}_4$ = significant difference
$\bar{X}_2$ vs $\bar{X}_3$ = significant difference
$\bar{X}_2$ vs $\bar{X}_4$ = significant difference
$\bar{X}_3$ vs $\bar{X}_4$ = not a significant difference

3. The group receiving no treatment allows the researcher to establish what changes in blood pressure might occur just with the passing of time. It serves as a measure of control.

<u>Post Test</u>

1. The null hypothesis is retained.
2. Variability between will exceed variability within
3. Treatment effect exists.
4. Through use of multiple comparison tests such as Scheffe's test.
5. Only when the null hypothesis is rejected.
6. Because all values are squared and thus positive.
7. All underlying populations are normally distributed and have equal variances.
8. No problem would exist as long as sample sizes are sufficiently large.
9. If sample sizes are equal, only one critical value must be calculated and all pairs may be compared using that value.

ANALYSIS OF VARIANCE (TWO WAY)

Learning Objectives

You should have knowledge and understanding of key terms.

You will be able to calculate and interpret the $\underline{F}$ test for two-way ANOVA.

You will be able to determine appropriate situations when the two-way ANOVA should be used.

Key Terms

Two-way ANOVA -- A more complex type of analysis of variance where population means differ with respect to two dimensions or factors.

Main effect -- The effect of a single factor when any other factor is ignored.

Simple effect -- The effect of one variable at a single level of another variable.

Interaction -- The product of inconsistent simple effects.

Text Review

In two-way ANOVA, there are four different types of means (refer to Table 18.1 in the text). The cell means referred to as (1)_____ (18.2) means reflect any effect due to the interaction of the two factors being studied. Column means represent the effect of one variable when the other is ignored. In the text example, column means represent reaction times for each number of confederates present when gender is ignored. Slight differences among these column means could be attributed to (2)_____ (18.2). More substantial differences reflect the main effect of the number of confederates on reaction time. In ANOVA, the effect of a single factor, when any other factor is ignored is the (3)_____ (18.2). The third type of mean, row means, represents the reaction times for gender when the number of confederates is ignored. Again, slight differences could be attributable to chance, but larger differences reflect (4)_____ (18.2) of gender on reaction time. The final average of the column means or row means equals the overall or (5)_____(18.1).

You may recall from Chapter 17 (p. 305, text) that in one-way ANOVA, a single F ratio is used to test the null hypothesis. In a two-way ANOVA, three different null hypotheses are tested, one at a time, with three F ratios: F_{column}, F_{row}, $F_{interaction}$. In each F ratio, the numerator represents (6)_____ (18.3). This variability can be either random error or random error plus (7)_____ (18.3). The denominator term represents only (8)_____ (18.3) for subjects treated similarly in the same group.

The hypothesis test for two-way ANOVA has three null hypotheses, each of which can be retained or rejected. The decision rule must mention the decision criteria and critical value for rejecting each hypothesis and each must be mentioned in the decision and interpretation. For a review, see the hypothesis test summary in your text.

The variance estimates in two-way ANOVA are similar to those in one-way ANOVA in that measures of variability always consist of a variance estimate or (9)_____ (18.4) calculated by dividing the sum of squares by its (10)_____ (18.4). As in one-way ANOVA, the major computational effort is in calculating the various sum of square terms. For computational formulas for the various SS terms and df terms, see Table 18.2.

The same procedures used for one-way ANOVA also apply to two-way for obtaining critical F values from Table C, Appendix D. With so many similarities between one-way and two-way ANOVA, (11)_____ (18.9) is the most obvious different feature of two-way ANOVA. Two factors are said to interact if the effects of one factor are not consistent for all the levels of the second factor.

Interaction can be clarified by examining the concept of simple effect, the effect of one variable at a single level of another variable. Interaction is the product of (12)_____ (18.9) simple effects. Whenever interaction is present, as indicated by the rejection of the null hypothesis for interaction, special statistical tests are usually conducted to pinpoint the precise

137

location of the discrepancies that caused it. These special tests are beyond the scope of this text but may be explored further in the Keppel book footnoted in your text. Differences between row means and column means may also be examined with a modification of Scheffe's test for multiple comparisons.

Assumptions for the F test in two-way ANOVA are similar to those in one-way ANOVA. The underlying populations are assumed to be (13) _____ (18.12) with equal variances. If sample sizes in all groups are equal and larger than (14) _____ (18.12), violations of these assumptions would be of small consequence. It is very important to have equal sample sizes in all groups. If this is not possible, consult a more advanced statistics book in regard to the issue of unequal sample size for ANOVA.

Although more complex ANOVA is possible with a larger number of factors, the goal of the researcher should be to use the simplest design that will adequately answer the research question.

Problems and Exercises

1. The example problem presenting one-way ANOVA in Chapter 17 (p. 307, text) involved using F to test the effect of the number of confederates present on sounding an alarm for smoke. Take this study further by adding the factor of whether the subjects grew up in rural or urban environments. Using the following data:

    a. Test the various null hypotheses at the .05 level of significance
    b. Summarize the results with an ANOVA table.

Type of Environment	Number of Confederates		
Rural	0	2	4
	15	8	10
	13	11	13
	10	7	11
	16	9	12
	14	11	9
Urban			
	4	11	15
	8	16	18
	5	10	19
	7	11	17
	3	12	19

2. A military psychologist wishes to determine whether men are really better than women at spatial relations tasks, whether they are trained or untrained recruits. The four groups of six randomly assigned men and women are administered a ten-item spatial relations test. The data follow.

   a. Test the various null hypotheses at the .01 level of significance.

   b. Summarize the results with an ANOVA table.

	Untrained Recruits	Training Recruits
Male	6	8
	6	6
	7	10
	4	5
	8	10
	8	9
Female	5	8
	5	6
	4	6
	3	6
	7	9
	3	5

Beyond the Basics

Reread Section 18.9 in your text describing interaction, paying close attention to the examples. Recall that interaction exists when the effects of one factor are not consistent for all levels of a second factor. Supply another example of factors that might interact from your own knowledge or experience. Try to determine how these factors could be set up for an experiment using ANOVA. Using the hypothesis test summary in your text as a model, write the research problem and statistical hypotheses. Set a level of significance and be prepared to defend your choice. Without any actual calculations (since you have no real data), predict an outcome and make a hypothetical decision to retain or reject the null hypotheses. Then write an interpretation based on your decision.

Post Test

1. In two-way ANOVA, if graphs are formed to reflect the possible effects, what is the meaning of slanted lines? of parallel lines?

2. In two-way ANOVA, what is represented by the numerator and denominator in the F ratio?

3. What is the basis for rejecting the null hypothesis in two-way ANOVA?

4. What is the most striking feature of two-way ANOVA?

5. What are the assumptions for the F tests in two-way ANOVA?

6. What happens if these assumptions are violated?

Answers

Text Review

1. treatment-combination
2. chance
3. main effect
4. a main effect
5. grand mean
6. variability
7. treatment effect
8. random error
9. mean square
10. degrees of freedom
11. interaction
12. inconsistent
13. normally distributed
14. ten

Problems and Exercises

1.

Source	SS	df	MS	F
Column	126.47	2	63.23	16.01
Row	1.20	1	1.20	0.31
Interaction	295.40	2	147.70	37.55
Within	94.4	24	3.93	
Total	517.47	29		

2.

Source	SS	df	MS	F
Column	20.37	1	20.17	7.16
Row	16.67	1	16.67	5.92
Interaction	.67	1	.67	.24
Within	56.33	20	2.82	
Total	93.83	23		

No interaction.

Post Test

1. Slanted lines indicate a possible main effect, parallel lines indicate a possible interaction.
2. The numerator represents variability between groups, columns, rows, and variability due to interaction. The denominator represents the variability within groups or random error.
3. A calculated value of F is compared to a critical value of F, and if the calculated value exceeds the critical value, the null hypothesis is rejected.
4. Interaction
5. The assumptions for the F test in two-way ANOVA are that underlying populations are normally distributed and have equal variances.
6. There is no concern for violations of these assumptions if sample size is larger than 10.

CHAPTER 19

CHI-SQUARE ($\underline{X}^2$) TEST FOR QUALITATIVE DATA

Learning Objectives

You should develop knowledge and understanding of key terms.

You will be able to determine appropriate situations for the use of chi-square for one- and two-variable cases.

You will be able to calculate and interpret chi-square tests.

Key Terms

One-way test -- Chi-square test that evaluates whether observed frequencies for a single qualitative variable are adequately described by hypothesized or expected frequencies.

Expected frequency -- The hypothesized frequency for each category, given that the null hypothesis is true.

Observed frequency -- The obtained frequency for each category.

Two-way test -- A chi-square test that evaluates whether observed frequencies reflect the independence of two qualitative variables.

Text Review

You may recall from Chapter 1, that when observations are classified into categories, the data are (1)_____
(19.0). The hypothesis test for qualitative data is known as chi-square. When the variables are classified along a single variable, the test is a one-way chi-square. The one-way chi-square test makes a statement about two or more population (2)_____
(19.1) that reflect the expected frequencies.

If the null hypothesis is true, then except for the effects of chance, the hypothesized proportions should be reflected in the sample.   The number of observations actually represented is referred to as (3)_____ (19.2) and is calculated bý multiplying the expected proportion by the total sample size.   If the discrepancies between the observed and expected frequencies are small enough to be attributed to chance, then the null hypothesis would be retained.   But if the discrepancies between the observed and expected frequencies are large enough to qualify as a rare outcome, the null hypothesis would be (4)_____ (19.2).

The value of chi-square can never be (5)_____ (19.3) because of the squaring of each difference between observed and expected frequencies.

For the one-way chi-square test, the degrees of freedom always equals the number of (6)_____ (19.4) minus one.

The chi-square test is nondirectional because the squaring of the discrepancies always produces a (7)_____ (19.6) value.   However, for the same reason, only the upper tail of the sampling distribution contains the rejection region.

It is possible to cross-classify observations along two qualitative variables.   This is referred to as a (8)_____ (19.6) chi-square test.   For the two-way test, the null hypothesis makes a statement about the lack of relationship between the two qualitative variablies.   In the two-way test, words are usually used instead of symbols in the null hypothesis, and as in the one-way test, the research hypothesis simply states that the null hypothesis is false.

In the two-way test, expected frequency is calculated by multiplying the column total times the row total and dividing by the overall total (see formula 19.4).   The chi-square critical value may be found in Table D of Appendix D only if degrees of freedom is known.   For the two-way test, degrees of freedom equals the number of categories for the column variable minus one, times the number of categories for the row variable minus one [df = (C - 1) (R - 1)].

Some precautions are necessary in using the chi-square tests.   One restriction is that the chi-square test requires that observations be (9)_____ (19.11). In this case, independence means that one observation should have no influence on another. One obvious violation of independence occurs when a single subject contributes more than one pair of observations. One way to check that this requirement is not being violated is to remember that the total for all observed frequencies must never exceed the total number of subjects. Using chi-square appropriately also requires that expected frequencies not be too small. Generally, any expected frequency of less than (10)_____ (19.11) is too small. Small sample sizes should also be avoided, as should unduly large sample sizes. A sample size that is too large produces a test that detects differences of no practical importance.

Problems and Exercises

1. The American Automobile Association believes that the three long weekends with Monday holidays are equally dangerous in terms of traffic fatalities.

   a. Using the .05 level of significance, test the null hypothesis for the following data.

   b. Specify the p-value for the test result.

Holiday Fatalities

Frequency	Presidents' Day	Memorial Day	Labor Day
f_o	347	396	379

2. A researcher takes a sample of 200 students at a small college. The variable of ethnicity is significant to the research such that it must be determined whether or not the sample differs from the underlying population. School records indicate the following breakdown of the general student body: 60% white, 22% black, 13% Chicano, 5% other. Using the following data from the research sample, test the null hypothesis at the .05 level of significance.

Ethnicity of College Students

White	Black	Chicano	Other
114	49	28	9

3. A sociologist examines the responses of 150 randomly selected people. The data are cross-classified on the basis of religious preference (Catholic/Non-Catholic) and attitude toward abortion in extenuating circumstances (rape, risk to life or health of Mother).

Catholicism and Attitude Toward Abortion

Religion	Attitude Toward Abortion		
	Favor	Oppose	Total
Non-Catholic	52	23	75
Catholic	39	36	75
Total	91	59	150

a. Using the .05 level of significance, test the null hypothesis that there is no relationship between religious preference and attitude toward abortion.

b. Speciy the p-value for the test result.

c. How might these results appear in a published report?

148

4.   A community college professor wanted to investigate whether traditional students (age twenty-three and younger) preferred instructional methods that were different from those preferred by nontraditional students (age twenty-four and older).

a.   Test the following results using the .05 level of significance.

b.   Specify the p-value for the test result.

c.   How might these results appear in a published report?

Instruction Methods

Student	Lecture,	Guest Speaker,	Media,	Group Study
Traditional	8	12	18	12
Nontraditional	17	11	14	8
Total	25	23	32	20

Beyond the Basics

The chi-square test represents a return to working with qualitative data. During the term, you may have thought of research questions dealing with qualitative data. If so, here is an opportunity to determine if chi-square would be an appropriate test to answer your question. If not, look back at the discussion of qualitative data in Chapter 1 (p.6, text). Are there variables there representing qualitative data that are of interest to you? Discuss with one of your classmates or your instructor what areas of the behavioral sciences or other sciences might be most likely to use the chi-square test. Do some fields of study seem more likely than others?

Which ones? Again, you might find more information by looking at professional journals in the library.

Post Test

1. When the researcher wishes to determine whether a population complies with a single set of hypothesized proportions, which chi-square test should be used?

2. When the researcher wishes to test whether there is a relationship between two qualitative variables, which chi-square test should be used?

3. Why is the chi-square test nondirectional?

4. What are the conditions for use of the chi-square test?

5. What is the difference between expected frequency and observed frequency?

Answers

Text Review

1. qualitative
2. proportions
3. expected frequency
4. rejected
5. negative

6. categories
7. positive
8. two-way
9. independent
10. five

Problems and Exercises

1. $\chi^2 = 3.31$, $\underline{df} = 2$, $\underline{p} > .05$, retain $\underline{H}_0$

2. $\chi^2 = 1.12$, $\underline{df} = 3$, $\underline{p} > .05$, retain $\underline{H}_0$

3. $\chi^2 = 4.90$, $\underline{df} = 1$, $\underline{p} < .05$, reject $\underline{H}_0$
   There is evidence that religious preference is related to attitude toward abortion [$\chi^2(1) = 4.90$, $\underline{p} < .05$].

4. $\chi^2 = 4.58$, $df = 3$, $\underline{p} > .05$, retain $\underline{H}_0$
   There is no evidence that type of student is related to preference for instructional method [$\chi^2(3) = 4.58$, $\underline{p} > .10$].

Post Test

1. One-way test
2. Two-way test
3. Because all discrepancies are squared so there can never be a negative value.
4. The conditions for using the chi-square test are (1) all observations must be independent, (2) expected frequencies must be sufficiently large, (3) sample size must be neither too small nor too large.
5. The expected frequency is the hypothesized frequency for a category that the researcher expects to obtain. The observed frequency is the actual observed frequency for each category.

TESTS FOR RANKED DATA

Learning Objectives

You should develop knowledge and understanding of key terms.

You will be able to determine appropriate situations for the use of tests for ranked data.

You will be able to calculate and interpret the various tests for ranked data.

Key Terms

Mann-Whitney U test -- A test for ranked data when there are two independent groups.

Wilcoxin T test -- A test of ranked data when there are two dependent groups.

Kruskal-Wallis H test -- A test for ranked data when there are more than two groups.

Nonparametric tests -- Tests, such as U, T, and H, that evaluate entire population distributions rather than specific population characteristics.

Distribution-free tests -- Tests, such as U, T, and H, that make no assumptions about the form of the population distribution.

Text Review

In this chapter, the U, T, and H tests are described. These tests are to be used with (1)_____ (20.0) data. Furthermore, these tests can be used when underlying populations can't be assumed to be normally distributed with equal variances.

The Mann-Whitney U test is used when there are two independent samples. The familiar t test cannot be used because the assumptions of normality and equal variances have been violated. This would affect the probability of a (2)_____ (20.1). The data are converted to ranks to avoid this problem. Using the ranks, the null hypothesis equates the two entire population (3)_____ (20.2). Therefore, any type of inequality between the population distributions could cause the rejection of the null hypothesis. If it can be assumed that the two population distributions have about equal variabilities and similar shapes, then rejecting the null hypothesis would indicate that the difference between the two is likely to be difference in central tendency or difference in population (4)_____ (20.2).

When all estimates in the two groups have been assigned ranks, the groups can be compared to form a preliminary impression. The more one group outranks the other, the larger the difference between the mean ranks for the two groups, the more likely the null hypothesis will be rejected. The calculated value for U is compared with the critical value found in Table E of Appendix D. The decision rule is unusual in that the null hypothesis will be rejected only if the observed U is less than or equal to the critical U. This is the opposite of the decision rule for all other tests presented so far in this text.

The U test can be either directional or nondirectional. However, when a directional test is desired, the researcher must meet the assumption of similar (5)_____ and (6)_____ (20.5). One caution that should be observed when using the directional test is that the researcher must make sure that differences in the population distributions are in the direction of concern.

If there are uncontrolled variables that could affect the outcome of the study, the subjects could be matched. The matching creates two dependent samples that require the Wilcoxin T test. The familiar t test cannot be used because the data in the text example are skewed, thus violating the assumption of (7)_____ (20.7) required for t. The null hypothesis for T is like that for U in that it equates the two (8)_____ (20.8). The rejection of the null hypothesis indicates only that the two populations differ. Again, if similar variabilities and shapes can be assumed, more precise conclusions are possible.

The observed value for T is compared to the critical T found in Table F of Appendix D. The T test may be either directional or (9)_____ (20.10). As with U, the null hypothesis will be rejected only if the observed $\underline{T}$ is (10)_____ (20.10) than or equal to the critical $\underline{T}$.

When there are three or more independent groups, the Kruskal-Wallis $\underline{H}$ test must be used if the assumptions of (11)_____ and equal (12)_____ (20.12) cannot be assured. In this test, unless sample sizes are very small, the critical values of $\underline{H}$ are obtained from the chi-square distribution (Table D, Appendix D). This requires degrees of freedom for $\underline{H}$, $\underline{df}$ = number of groups - 1. Unlike $\underline{U}$ and $\underline{T}$, the decision rule for $\underline{H}$ returns to the more familiar pattern. The null hypothesis will be rejected if the observed $\underline{H}$ is equal to or greater than the critical chi-square. Because the sums of the ranks are squared, the $\underline{H}$ test is always (13)_____ (20.17).

The $\underline{U}$, $\underline{T}$, and $\underline{H}$ tests, as well as the chi-square test, are referred to as nonparametric tests. Parameter refers to any descriptive measure of a population, such as a population mean. Nonparametric tests evaluate hypotheses for entire population distributions. The parametric tests like $\underline{t}$ and $\underline{F}$ evaluate hypotheses for a specific parameter, usually the population mean. Nonparametric tests may also be referred to as (14)_____ (20.19). The name signifies that these tests require no assumptions about the form of the population distribution. Remember, you have learned that $\underline{t}$ and $\underline{F}$ tests require that populations be normally distributed and have equal variances. The $\underline{U}$, $\underline{T}$, and $\underline{H}$ tests make no such requirements.

When data are ranked and the data are quantitative but don't seem to originate from normally distributed populations with equal variances, use (15)_____, _____, and _____ (20.20) tests. When the data are quantitative and the populations appear to be normally distributed with equal variances, use (16)_____ and _____ tests (20.20). Under the appropriate assumptions, $\underline{t}$ and $\underline{F}$ are more likely to detect a false null hypothesis, thus minimizing the chance of a type II error.

Problems and Exercises

1. Can parental training in behavior modification techniques improve first graders' classroom behavior? Sixteen first graders were randomly assigned to two groups, one whose parents received behavior modification training and one whose parents did not. Two months after the training, the first-grade teacher was asked to rank the sixteen children on compliance to teacher requests.

    ### Compliance Ranking of First Graders

Parents with Training	Parents without Training
1	11
3.5	14.5
9	16
13	10
3.5	14.5
7	6
2	8
5	12

    a. Use U to test the null hypothesis at the .05 level of significance.

    b. Specify the p-value for the test result.

    c. Interpret your results.

2. A team of psychologists worked with students on the problem of test anxiety. Using the Swinn Test Anxiety Behavior Scale, test anxiety is measured before and after a series of group workshops and individual therapy to reduce anxiety. The following results were obtained:

Swinn Test Anxiety Behavior Scale Scores

Student	Before	After
1	120	105
2	135	100
3	145	115
4	160	125
5	120	110
6	150	120
7	135	110

a. Use T to test the null hypothesis at the .05 level of significance.

b. Specify the p-value for the test result.

c. Interpret your results.

3. The same consumers' group mentioned in problem 6 in the text continues its interest in motion picture ratings. After the films were screened for violence and sexually explicit scenes, an additional viewing was done by the trained observer to count the number of expletives in each film. The following results were obtained.

Number of Expletives in Films

X	R	PG 13	PG	G
12	17	15	10	2
8	22	13	7	0
7	19	11	5	3
10	19	17	8	3
7	21	15	7	1

a. Use H to test the null hypothesis at the .05 level of significance.

b. Specify the p-value for the test result.

c. Interpret your results.

4. The principal of an elementary school wishes to determine the most effective method to improve reading as measured by the district's reading achievement test. Thirty children in third grade are randomly assigned to one of three reading improvement programs. Test the difference between their September reading scores and April reading scores to determine whether any one method is superior. Use the .05 level of significance. The difference scores follow.

You may recognize this problem from Chapter 17 (p. 128, workbook).

a. Instead of the F test, just for practice, test the null hypothesis using the H test.

b. Specify the p-value for the test result.

c. Interpret your results.

Tutoring	Home Help	Reading Aloud Practice
10	8	6
7	6	9
12	4	5
9	4	7
13	9	7
11	7	6
10	8	8
14	5	5
11	7	9
12	8	4

Beyond the Basics

To help you really understand the nonparametric tests, make a chart comparing each of the nonparametric tests with its parametric counterpart. For example, compare t with U. You would want to include information such as the assumptions for each test, the type of data appropriate for use with the test, how to determine critical values, and how to make the decision to retain or reject H_0.

Post Test

1. What is the appropriate test for ranked data with two independent samples?

2. What test is appropriate for ranked data using two dependent samples?

3. Why are these tests used instead of t?

4. What is different about the decision rule for the Mann Whitney U test and the Wilcoxin T test?

5. What is a nonparametric test?

159

6. What is a distribution-free test?

7. Which of the tests studied in this chapter seems to replace ANOVA under specific circumstances?

Answers

Text Review

1.	ranked	9.	nondirectional
2.	type I error	10.	less
3.	distributions	11.	normality
4.	means	12.	variances
5.	variabilities	13.	nondirectional
6.	shapes	14.	distribution-free
7.	normality	15.	U, T, H
8.	entire populations	16.	t, F

Problems and Exercises

1. Using a one-tailed test, U = 8, reject H_0, $p < .01$
   There is evidence that parental training in behavior modification influences first graders' compliance with teacher requests.
2. Using a one-tailed test,
   T = 0, reject H_0, $p < .01$
   The evidence suggests that therapy and workshops helped reduce anxiety.
3. H = 21.877, df = 4, reject H_0, $p < .05$
   The evidence suggests that movies with different ratings contain different numbers of expletives.
4. H = 15.697, df = 2, reject H_0, $p < .05$
   The evidence suggests that the reading methods differ in improving the reading of third graders.

160

Post Test

1. Mann-Whitney U test
2. Wilcoxin T test
3. These tests must be used when the data are ranked or the assumptions of normality and equal variances in the populations are violated.
4. In the decision rule for each of these tests, the null hypothesis is rejected if the observed value is less than or equal to the critical value.
5. Nonparametric tests are those such as U, T, and H, that evaluate entire population distributions rather than specific population characteristics.
6. Distribution-free tests are those, such as U, T, and H, that make no assumptions about the form of the population distribution.
7. Kruskal-Wallis H test

POSTSCRIPT: WHICH TEST?

Learning Objectives

You will be able to determine which statistical analysis is appropriate to answer a particular research question.

Text Review

If you are attempting a research project of your own, you will probably be selecting a statistical test from among the ones you have studied so far.   The following is a review of some important ideas and a look at some questions you might ask in order to select the appropriate test.

What is the intent of your study?   If you wish to summarize existing data, you will be using (1)_____ statistics (21.1).   If you wish to generalize beyond existing data, you will be using (2)_____ (21.1) statistics.

In the behavioral sciences, hypothesis tests are usually preferred to confidence intervals.   However, even if you follow this preference, if the null hypothesis is rejected, consider estimating the possible size of the effect by constructing a (3)_____ (21.2).

Another question that must be answered when deciding which test to use is whether the observations are quantitative or qualitative.   If the data are (4)_____, (21.3) the appropriate hypothesis test should be selected from the various t or F tests or their nonparametric counterparts.   When the observations are (5)_____, (21.3) the appropriate test will be the chi-square.

Sometimes the decision as to whether the data are qualitative or quantitative is not so easy.   If you are not sure, use the following guidelines to help make your decision.   (1)   Focus on a single observation.   (2) Focus on numerical summaries.   (3)   Focus on key words.

Once the nature of the data has been decided, another question to be resolved in selecting the appropriate test is the number of groups.   If only one group is involved, then a (6)_____ (21.5) test for a single population is appropriate. If there are two groups, a t test may also be appropriate, but you must decide whether the groups are dependent (paired or matched) or independent. If the concern of the researcher is to determine whether paired observations are significantly correlated, then the appropriate test is the t test for a population correlation coefficient. When there are more than two groups, the F test for (7)_____ (21.5) is the correct choice. The ANOVA test can be either one-way or (8)_____ (21.5).

Finally, the nonparametric tests are to be used when the original observations are (9)_____ (21.6) or when some assumption is violated.

Problems and Exercises

Here are some additional exercises to provide practice in identifying the appropriate statistical test. As with the exercises in the text, no assumptions have been violated unless noted otherwise. For each example, be sure to specify all you know about the type of test. For instance, specify that the t test is for a single population mean, or that chi-square is one-way or two-way.

1. A researcher wishes to determine the relationship between scores on the GED test and scores on the ACT.

2. An investigator wishes to determine whether drinking alcohol impairs memory. Before a test of short-term memory, subjects drank a glass of juice. For one group, the juice was straight and for the other group, the juice contained 2 ounces of vodka. The volunteer subjects were randomly assigned to the two groups. The mean performance was computed for each group.

3. The preceding experiment was repeated using the same subjects for both conditions. On Monday half the subjects drank the plain juice and performed a short-term memory test and the other half drank the juice with the alcohol and then performed the memory test. On the following day, the procedure was

163

repeated according to counterbalancing requirements. The mean performances were computed.

4. Do male and female college students desire different characteristics in a spouse? The students are surveyed as to what characteristics they most desire. Responses of male and female students indicate that sense of humor, looks, loyalty, wealth, and educational level are important to both sexes, but not necessarily in equal proportion.

5. A developmental psychologist wishes to examine the relationship between age and reaction time. Three groups of subjects are tested on a measure of reaction time and their performances are compared. Subjects in group 1 are ten years old. Group 2 subjects are forty years old, and group 3 subjects are seventy years old.

6. A psychobiologist wishes to test the visual performance of kittens after varying periods of light deprivation. At birth, the kittens are randomly assigned to periods of zero, one, two, or three months of light deprivation. The visual performance is tested by a trained observer who rates the kittens' performance in a maze that produces a challenging visual task. A mean performance is computed for each group.

7. A psychology professor wishes to determine whether the test performance of statistics students is affected by special tutoring after class. The students are randomly assigned either to a tutoring group or to a group that meets after class socially but does not study or discuss course material. The test performances of group members are ranked.

8. The statistics students are randomly assigned to a test anxiety workshop in a group or to individual therapy to reduce the test anxiety. The researcher wishes to determine which method best reduces the test anxiety scores.

9. A researcher wishes to determine the relationship between scores on a love scale and number of years of marriage.

10. A director of institutional research wishes to determine whether students' math exam grades are higher when math classes are taught twice a week for one and a half hours, once a week for three hours, or three times a week for fifty minutes.

Answers

Text Review
  1. descriptive
  2. inferential
  3. confidence interval
  4. quantitative
  5. qualitative
  6. t
  7. ANOVA
  8. two-way
  9. ranked

Problems and Exercises

  1. Correlation - Pearson r
  2. t test for independent samples
  3. t test for dependent samples
  4. two-way chi-square
  5. one-way F
  6. one-way F
  7. Mann-Whitney U test
  8. t test for independent samples
  9. Correlation - Pearson r
 10. one-way F